Municipal Sludge Use in Land Reclamation

William E. Sopper
School of Forest Resources
The Pennsylvania State University

LEWIS PUBLISHERS
Boca Raton   Ann Arbor   London   Tokyo

Library of Congress Cataloging-in-Publication Data

Sopper, William E.
   Municipal sludge use in land reclamation / William E. Sopper.
      p.   cm.
   Includes bibliographical references and index.
   ISBN 0-87371-941-7
   1. Mineral industries—Environmental aspects.   2. Sewage sludge as
fertilizer.   3. Reclamation of land.   I. Title.
TD195.M5S64   1993
627'.5—dc20   92-41641
                                                                        CIP

Direct all inquiries to CRC Press, Inc., 2000 Corporate Blvd., N.W., Boca Raton,
Florida 33431.

PRINTED IN THE UNITED STATES OF AMERICA
1 2 3 4 5 6 7 8 9 0
Printed on acid-free paper

Preface

The disposal of municipal sewage sludge is an environmentally sensitive problem facing municipalities throughout the world. As environmental quality standards become more stringent and volumes of sludge generated continues to increase, municipalities are confronted with an urgent need to develop safe and feasible alternative practices for sludge management. Traditional sludge disposal methods are coming under increasing pressure to change. Ocean dumping is being phased out, incineration is costly and contributes to air pollution, and landfill space is becoming scarce. One possible long-term solution appears to be recycling of the sludge and using it for beneficial purposes. Land application of sludge on agricultural lands, forest lands, and disturbed lands represents the best option for utilization of this material. Unfortunately, the general public often opposes these practices because of their concerns over adverse environmental effects, such as soil contamination, groundwater pollution, and the potential threat to the human and animal health. Much research has been done over the past 25 years that shows that many of these concerns are unwarranted. Mining of all types has disturbed millions of hectares throughout the world. Many of these hectares have never been properly restored to their original condition. These naked lands represent a great potential for the utilization of large volumes of sludge. Most of these lands are difficult to revegetate because they have a low pH, are devoid of nutrients and organic matter, have a low water-holding capacity, and may contain toxic concentrations of trace metals. Applications of sludge can ameliorate these conditions and facilitate the rapid establishment of a permanent vegetative cover.

This book has been an attempt to review all of the available literature, on an international scope, related to the use of municipal sludge to reclaim disturbed land and to evaluate the potential effects on the environment. The general conclusion was that stabilized municipal sludges, if applied properly according to present guidelines and regulations, can be used to revegetate disturbed lands in an environmentally safe manner with no major adverse effects on the vegetation, soil, or groundwater quality and does not pose any significant threat to animal or human health.

The preparation of this book manuscript was funded through a contract from the Office of Wastewater Enforcement and Compliance, U.S. Environmental Protection Agency, Washington, DC, Robert K. Bastian, project officer.

This book should be of interest to researchers and practitioners working in the fields of mine land reclamation and land application of waste materials. It should be useful to consulting firms and businesses involved in land application of municipal sludge. It should also be useful to local, state, and federal officials involved in the development of guidelines and regulations for the beneficial uses of municipal sludge. The book would also be a useful reference for faculty and graduate students for college courses related to mine land reclamation, land application of wastes, and environmental resource management.

William E. Sopper

Dr. William E. Sopper received his B.S. and M.F. in forestry at the Pennsylvania State University. He continued his graduate work in forest hydrology at Yale University and received his Ph.D. in 1960. He is currently a Professor of Forest Hydrology in the School of Forestry and is Co-Director of the Center for Mine Land Reclamation in the Environmental Resources Research Institute at Penn State. He has received several professional awards, the most recent being the 1992 William T. Plass award given by the American Society for Surface Mining and Reclamation for his preeminent research career in mined land reclamation. He is the authors of over 200 published articles related to the land application of municipal wastewater and sludge and the revegetation of mined land. He is co-editor of four books, *Forest Hydrology* published in 1967, *Recycling Treated Municipal Wastewater and Sludge through Forest and Cropland* (1973), *Utilization of Municipal Sewage Effluent and Sludge on Forest and Disturbed Land* (1979), and *Land Reclamation and Biomass Production with Municipal Wastewater and Sludge* (1982).

Table of Contents

1. Introduction 1
   Status of Land Disturbed by Mining 1
   Surface Mining Control and Reclamation Act of 1977 6
   Federal and State Regulations Governing Use of Sludge on
   Mine Land 8

2. Review of Land Reclamation Projects Using Municipal Sludge 13
   Overview 13
   Effects on Vegetation 13
       Growth Responses 13
           Grass and Legume Species 13
           Field Crops 37
           Trees 41
           Summary 47
       Vegetation Quality 48
           Macronutrients 48
           Trace Metals 52
               Grass and Legume Species 52
               Trees 80
               Field Crops 88
           Summary 90
   Effects on Soil 91
       Physical Properties 91
           Summary 92
       Chemical Properties 92
           Summary 103
       Biological Properties 103
           Aerobic Heterotrophic Bacteria 106
           Fungi 106
           Actinomycetes 108
           Nitrifying Bacteria 108
           Soil Community Respiration 109
           Microbial Decomposition 110
           Trace Metals 112
           Organic Matter 114
           Nitrogen Mineralization.. 124
           Earthworms 125
           Summary 125
   Effects on Water Quality 126
       Soil Water and Groundwater 126
           Nitrate–Nitrogen 126
           Trace Metals 128
       Surface Water 131

Summary 132

Effects on Animal Nutrition and Health 132

Summary 139

U.S. EPA 40 CFR Part 503 Final Rules for Use and Disposal of
Sewage Sludge 139

3. Conclusion 141

4. References 143

5. Appendix 157

Index 159

CHAPTER 1

Introduction

STATUS OF LAND DISTURBED BY MINING

Disturbed land resulting from both surface and underground mining can result in major water quality problems as well as being unsightly and unproductive. The largest acreages of disturbed land are created by coal mining activities. The major coal fields in the United States are shown in Figure 1. The U.S. mining industry has disturbed over 1.48 million ha between 1930 and 1971, and only 40% of this has ever been reclaimed (Paone et al., 1978). Besides coal, mining for sand, gravel, stone, clay, copper, iron ore, phosphate rock, and other minerals account for most of the disturbed land. Table 1 shows the status of lands in the U.S. disturbed by surface mining, including both land requiring reclamation by law and abandoned mine lands for which there is no legal requirement for reclamation. Pennsylvania, Ohio, Kentucky, and Illinois each have over 64,000 ha and West Virginia, Alabama, and Missouri over 29,000 ha of unreclaimed coal-mined land. Florida has more than 103,000 ha of unreclaimed land after phosphate and other mining activities.

Surface mining, half of which is for coal, has disturbed over 1.6 million ha in the U.S. Additional thousands of hectares will be disturbed each year (Table 2).

The Office of Surface Mining Reclamation and Enforcement (1987) recently released a ten-year progress report on mine land reclamation since the passage of the Surface Mining Control and Reclamation Act of 1977. The amount of hectares under permit for surface, underground, and other mining operations during the period 1977 to 1986 is given in Table 3. The amount of land under permit peaked in 1984 at 265,768 ha and then declined to 107,429 in 1986. The states with the largest amount of land under permit were Kentucky (3,489 ha), Pennsylvania (1,505 ha), and West Virginia (1,443 ha). The amount of land reclaimed with bonds released during the same period is given in Table 4. The amount of land successfully reclaimed represents approximately 41% of the land under permit,

Figure 1. Coal fields of the United States.

showing that considerable progress is being made toward revegetation of land disturbed by mining activities.

Although most legislative and technical experiences in land reclamation have involved land surface mined for coal, the information may be applied, with modifications, to other types of disturbed or marginal land. Forestry-related activities, roadway and other construction, and deposition of dredge materials and fly ash from coal-fired power plants create large tracts of wasteland which are often difficult to reclaim with conventional techniques and offer considerable potential for municipal sludge amendments. Large acreages of forest are harvested or devastated by forest fires, landslides, and other natural disasters each year which require reestablishment of trees for return to productivity.

About 400 million m³ of sediment are dredged each year in the maintenance and establishment of waterways and harbors. Many sites where dredge spoils are deposited are highly acidic and of low productivity. Eroding sediment pollutes nearby waterways. Municipal sludge has been successfully used to stabilize and revegetate acidic dredge spoils along the Chesapeake and Delaware Canal (Palazzo, 1977).

Nearly 700 million metric tons of fly ash, cinders, and bottom ash from coal-fired power plants have been produced since the end of World War II (Palazzo, 1977), and with the increasing construction of new power plants, it is estimated that about 70 million metric tons of ash and 10 million metric tons of flue gas desulfurization sludge per year (U.S. EPA, 1979) will be produced. Fly ash contains some essential plant nutrients, and studies have evaluated the waste material in combination with municipal sludge for reclaiming derelict areas (Sutton, 1980; Cavey, 1980).

Another area where sludge could be beneficially used is in stabilizing/revegetating construction sites and roadways. Over 10 million ha are occupied by public roads and highways in the U.S. Some work has been done using sludge to stabilize and revegetate these often eroded and unproductive soils (Gaskin et al., 1977;

Table 1.   Status of Land Disturbed by Surface Mining in the United States as of July 1, 1977

| State | Reclamation not Required by any Law (ha) | | | Reclamation Required by Law (ha) |
	Coal Mines	Sand and Gravel Pits	Other Mine Areas	All Disturbed Lands
Alabama	29,278	8,954	10,603	18,856
Alaska[a]	1,093	1,741	1,620	—
Arizona[a]	162	2,592	24,664	—
Arkansas	2,277	8,700	4,649	1,811
California	4	3,228	32,804	28,130
Colorado	2,871	3,375	6,424	7,848
Connecticut[a]	—	6,780	319	—
Delaware	1,179	25	609	—
Florida	—	5,883	103,932	9,836
Georgia	990	3,230	15,301	7,759
Hawaii	—	6	47	—
Idaho	—	2,065	607	8,788
Illinois	64,642	11,709	7,593	21,885
Indiana	40,688	6,501	3,408	32,664
Iowa	5,669	4,109	2,600	7,466
Kansas	16,709	4,515	4,114	3,414
Kentucky	103,621	1,328	3,034	64,515
Louisiana[a]	—	15,116	1,032	—
Maine	—	12,606	1,214	1,302
Maryland	4,906	6,954	1,180	6,957
Massachusetts[a]	—	12,977	4,184	—
Michigan	57	22,310	11,135	7,992
Minnesota	—	12,169	18,144	8,236
Mississippi	—	18,616	3,168	—
Missouri	32,181	2,235	13,868	6,429
Montana	792	1,885	7,428	6,422
Nebraska[a]	—	7,277	1,632	—
Nevada[a]	—	495	1,035	—
New Hampshire[a]	—	5,154	169	—
New Jersey[a]	—	9,967	2,256	—
New Mexico	9	4,803	731	12,489
New York	—	18,993	9,837	8,511
North Carolina	—	7,697	3,524	4,457
North Dakota	425	814	81	2,734
Ohio	100,872	15,908	11,077	4,137
Oklahoma	14,628	2,697	5,712	5,335
Oregon	—	1,426	7,115	3,384
Pennsylvania	121,500	10,530	18,427	40,500
Rhode Island[a]	—	1,050	—	—
South Carolina	—	5,451	2,155	3,073
South Dakota	360	4,112	2,130	3,046
Tennessee	13,247	2,333	1,393	2,054
Texas	1,340	61,745	15,027	6,076
Utah	257	1,620	1,788	6,069
Vermont	—	1,723	866	177
Virginia	12,938	3,125	1,318	5,732
Washington	19	3,929	3,310	5,704
West Virginia	37,473	1,844	403	3,102
Wisconsin	—	21,664	4,220	5,973
Wyoming	3,911	1,487	5,012	33,404

Source: U.S. Department of Agriculture (1980).

[a] No state law when surveyed; therefore, no reclamation by law.

Table 2.   Projected Regional Land Use for Coal Production from Surface Mining

Region	1975	1977	1980 (ha)	1985	1990
Northern Appalachia	8,019	8,302	9,396	10,813	14,013
Southern Appalachia	5,508	5,710	6,804	8,505	10,287
Midwest	7,209	7,411	8,302	9,922	11,704
Gulf	810	1,417	3,888	5,791	7,209
Northern Great Plains	400	500	600	900	1,100
Rocky Mountain	500	600	600	700	900
Pacific Coast	600	800	1,200	1,900	2,200
Total[a]	23,000	25,000	31,000	39,000	47,000

Source: Paone et al., (1978).

[a]   Data may not add to totals because of independent rounding.

Table 3.   Hectares Under Permit for Surface, Underground, and Other Mining Operations for Period 1977 to 1986[a]

Year	Total Hectares
1978	68,635
1979	146,002
1980	141,842
1981	153,880
1982	136,568
1983	182,896
1984	265,768
1985	175,057
1986	107,429
Total	1,378,077

Source: Adapted from Office of Surface Mining Reclamation and Enforcement (1987).

[a]   States and Indian tribe lands included in above tabulations were Alabama, Alaska, Arkansas, Illinois, Indiana, Iowa, Kansas, Kentucky, Louisiana, Maryland, Missouri, Montana, New Mexico, North Dakota, Ohio, Pennsylvania, Tennessee, Texas, Utah, Virginia, Washington, West Virginia, Wyoming, Crow Tribe, Hopi Tribe, and Navajo Tribe.

Palazzo et al., 1980), and sludge appears to have potential as an amendment along roads and right of ways.

Other types of drastically disturbed land where sewage sludge might be used beneficially are phosphate mines, copper mines, oil shale mining, zinc and lead smelters, quarries, sand and gravel pits, borrow pits, and landfills. The United States has four major phosphate rock-producing areas: (1) Florida, (2) North Carolina, (3) Tennessee, and (4) the western States of Idaho, Montana, Utah, and Wyoming. Trends in United States production have been forecast by the Bureau of Mines, U.S. Department of Interior (Stowasser, 1975) and indicate that phosphate

Table 4.   Number of Hectares Reclaimed
with Bonds Released During the
Period 1977 to 1986[a]

Year	Total Hectares
1978	19,078
1979	42,580
1980	51,401
1981	52,547
1982	80,351
1983	69,874
1984	96,910
1985	81,696
1986	68,280
Total	562,717

Source: Adapted from Office of Surface Mining
Reclamation and Enforcement (1987).

[a]   States and Indian tribe land included in above
tabulation are the same as Table 3.

production in the western area and North Carolina will probably continue to increase up to year 2015. Most of the oil shale deposits in the U.S. are located in Colorado, Utah, and Wyoming. Mining for oil shale produces several kinds of waste material and different reclamation techniques may be required for each type. The most difficult waste disposal problem and revegetation problem will be associated with the spent shale from the surface retorting operation. Waste rock removed to reach oil shale and low-grade shale which cannot be retorted profitably will also probably be surface stockpiled and will require stabilization and revegetation.

Deep coal mining refuse banks are another type of made-man disturbed land that represents a reclamation challenge. Abandoned coal refuse banks dot the landscape throughout the northern Appalachian region in the U.S. Mining of coal brings to the surface enormous amounts of black, shaley, acidic refuse material which was traditionally deposited in high conical-shaped banks. Thousands of hectares of such material, produced by over a century of mining, were left unreclaimed prior to the federal reclamation act of 1977. Most of these banks have remained barren and defy revegetation by natural processes. Side slopes of these banks are usually very steep and highly susceptible to severe erosion. The refuse material is black in color, resulting in extremely high surface temperatures during the summer growing season. The material is low in nutrients, has a low water-holding capacity, and generally has a pH lower than optimum for plant growth. The refuse banks are not only unsightly and unstable but also pose a threat to health and safety. They are a constant source of dust during the summer which often coats nearby houses and aggravates the health problems of persons with asthma, allergies, and other breathing ailments. The current method used to revegetate steep coal waste banks involves topsoiling followed by hydroseeding. This often requires the transportation of large volumes of overburden material from an active mine to the abandoned coal waste bank. This is a costly operation and often only fairly successful. Regrading waste banks is also very costly and in many cases space is not available for extensive leveling. Some coal waste banks in Pennsylvania are completely surrounded by occupied houses. Thus, what is needed is a technique to revegetate and stabilize these banks in situ. The use of sewage sludge as a substitute for topsoiling is one possibility.

Emissions from zinc and lead smelters have devastated thousands of hectares of land around the world. The New Jersey Zinc Company smelter at Palmerton, PA is a prime example of such devastation. Operating since 1898, emissions of Zn, Cd, Cu, Pb, and SO_4 have resulted in over 800 ha of what might be described as a biological desert on the north slope of Blue Mountain. Approximately 50% of the slope area is strewn with rocks and boulders exposed by severe erosion over the past 50 years, which has removed approximately 30 cm of the original surface soil. Natural revegetation has been hampered primarily by the high Zn concentrations in the surface mineral soil and organic detritus material present on the area. Soil nutrients have been washed away, microorganism populations are virtually nonexistent, and the surface soils are droughty with extreme variations in microclimate. As a result, the normal balance of vegetation, soil microorganisms, and the recycling of soil nutrients have been severely disrupted by the accelerating denudation and erosion.

The Blue Mountain site represents a unique set of reclamation challenges. First, the area is totally inaccessible to vehicles because the slopes are covered by rocks, boulders, and undecomposed tree stems. Access would only be possible by bulldozing new roads. The slopes are steep, averaging 30% and ranging from 25 to 100%. Most vehicles used to spread lime, fertilizer, or sludge must be able to traverse the site. This is not possible on this site, so it would be necessary to use a spreading vehicle that could apply the amendments aerially over considerable distances (30 to 45 m) onto the slopes. Incorporation of the amendments, a usual practice, would also not be possible on this site. And lastly, the surface soil is highly contaminated with trace metals, providing another impediment to vegetation establishment. Background surface soil samples taken downwind approximately at midslope on Blue Mountain contained 31,316 ppm of Zn, 5,225 ppm of Pb, and 1,248 ppm of Cd (Shoener et al., 1987). The normal ranges for uncontaminated soils in the United States are 10 to 300 ppm for Zn, 2 to 200 ppm for Pb, and 0.01 to 7.00 ppm for Cd (Allaway, 1968). A considerable portion of the area is also covered with a layer of black humus detritus that appears to be undecomposed tree bark. Analyses of this material showed equally high concentrations of trace metals — 29,475 ppm of Zn, 1,221 ppm of Cd, and 6,237 ppm of Pb. Soil samples taken to a depth of 90 cm indicated that the depth of adverse contamination for Zn (>200 ppm) and Cd (>5 ppm) ranged from 5 to 30 cm. No known efforts at reclaiming such a site have ever been successful. Sites such as this represent the ultimate in reclamation challenges.

SURFACE MINING CONTROL AND RECLAMATION ACT OF 1977

Revegetation of disturbed lands is currently an area of environmental concern and active research, as well as practical application, particularly since the 1977 federal Surface Mining Control and Reclamation Act established strict regulations for the revegetation of currently mined land. The act (PL 95-87, Section 515) requires that a diverse, effective, and permanent vegetative cover of the same seasonal variety native to the area of land to be affected must be established and must be capable of self-regeneration and plant succession at least equal in extent of cover to the natural vegetation of the area. In March 1982 amended regulations were published. These rules are currently set forth in 30 CFR 816 and 817 (Federal Register, 1982). Specifically, they state that:

1. The permanent vegetative cover of the area must be at least equal in extent of cover to the natural vegetation of the area and must achieve productivity levels compatible with the approved postmining land use. Both native and introduced vegetation species may be used.
2. The period of responsibility initiates after the last year of augmented seeding, fertilizing, irrigation, or other work which ensures revegetation success.
3. In areas of more than 66 cm of average annual precipitation, the period of extended responsibility will continue for not less than five years. In areas with 66 cm of precipitation or less, the period of responsibility will continue for not less than ten years.
4. Normal husbandry practices essential for plant establishment would be permitted during the period of responsibility so long as they can reasonably be expected to continue after bond release.
5. In areas of more than 66 cm of precipitation, the vegetative cover shall be equal to the success standard only during the growing season of the last year of the responsibility period unless two years would be required by the regulatory authority. In areas with less than 66 cm, the vegetative cover must be equal to the success standard for the last two years of the responsibility period.
6. The ground cover, productivity, or tree stocking of the revegetated area shall be considered equal to the success standard approved by the regulatory authority when the parameters are fully equivalent with 90% statistical confidence.

It will be difficult to meet these requirements using current reclamation techniques. New methods will have to be developed and larger amounts of lime, fertilizer, and seed will undoubtedly be needed. Soil amendments, mulching, and even irrigation may be required on some sites.

In addition to land disturbed by coal mining, other areas continually in need of reclamation in the U.S. include borrow pits, dredge spoils, construction sites, quarries, gravel pits, clear-cut and burned forests, and shifting sand dunes.

The problem of disposing of ever-increasing amounts of municipal sewage sludge is one which must be faced by many U.S. municipalities, particularly with the increasing cost of incineration, the decreasing land available for landfills, and the current controversy and concern over ocean dumping. It is estimated that more than 7.7 million dry tons of municipal sludge are currently produced each year by the 15,300 public-owned treatment plants in the U.S. Approximately 25% of this is being land applied for its fertilizer and organic matter value (Federal Register, 1989).

Both problems, that of devastated lands and that of sludge disposal, may be alleviated by sludge utilization and recycling, i.e., using sewage sludge to aid revegetation.

During the past two decades, extensive research has been carried out in the U.S. on the feasibility of reclaiming disturbed land with sludge. The research, as well as large-scale practical projects and commercial ventures, have shown that stabilized municipal sludge is an excellent soil amendment and chemical fertilizer substitute. Consequently, there has been considerable use of sludge for the production of agricultural crops. One disadvantage, however, is that sludge may contain every element or compound found in wastes from domestic and industrial sources. Thus, some concern has been raised about the potential introduction of these elements, particularly heavy metals, into the human food chain. The U.S. Environmental Protection Agency as well as some states have developed guidelines and regulations governing sludge applications in agriculture (Bastian et al., 1982).

FEDERAL AND STATE REGULATIONS GOVERNING USE OF SLUDGE ON MINE LAND

Most of these guidelines set limits on sludge application rates based on nitrogen and other nutrient requirements of the vegetation as well as trace metal loadings. For instance, the U.S. Environmental Protection Agency has developed guidelines concerning the maximum amounts of Pb, Zn, Cu, Ni, and Cd allowable on agricultural land used for growing food-chain crops (U.S. EPA, 1977). Food-chain crops are typically defined as those crops that can enter the human diet either with (wheat, corn) or without (leafy vegetables) processing. Researchers in the USDA and Agricultural Experiment Stations have proposed similar trace metal limits which would allow the growth of all crops after termination of sludge applications, provided the soil pH is maintained at 6.5 or above (Knezek and Miller, 1976, Baker et al., 1985). However, no federal guidelines have been issued specifically governing the use of sludge for reclamation purposes. Although the above guidelines were developed for agricultural applications, it is suggested that they also be considered for reclamation applications unless there are more specific state regulations.

The metal loadings suggested by the U.S. Environmental Protection Agency are given in Table 5. They are based upon the soil cation exchange capacity (CEC). The use of soil CEC was based on the assumption that metal solubility and thus plant availability tend to decrease with increasing CEC in most soils of the U.S.

These are *guidelines* and not *regulations*. The U.S. Environmental Protection Agency has only issued *regulations* for cadmium as part of the requirements under the Resource Conservation and Recovery Act of 1976 and the Clean Water Act of 1977 (Federal Register, 1979). These regulations are specifically for cadmium additions to cropland and can be summarized as follows:

1. The pH of the soil must be ≥ 6.5 at the time of sludge application.
2. Annual cadmium additions are limited to 0.5 kg ha^{-1} yr^{-1} if leafy vegetables or tobacco are grown.
3. For other food-chain crops, the annual cadmium additions cannot exceed 0.5 kg ha^{-1} yr^{-1}.

Table 5.   U.S. EPA Recommended Maximum Amounts of Trace Metal Loadings
             for Agricultural Cropland

	Soil Cation Exchange Capacity (c mol kg^{-1})		
	<5	5–15	>15
Metal	Amount of metal (kg ha^{-1})		
Pb	560	800	2240
Zn	280	560	1120
Cu	140	280	560
Ni	140	280	560
Cd	6	11	22

Source: U.S. EPA (1983).

4. The cumulative cadmium applied must be <5 kg ha^{-1} if the background soil pH is ≤6.5.
5. The cumulative cadmium applied is as shown in Table 5 for soils with a background pH ≥6.5 and for soils with a background pH ≤6.5 provided the pH is 6.5 at the time food-chain crops are grown.

The U.S. Environmental Protection Agency, Food and Drug Administration, and the U.S. Department of Agriculture have also recommended that cumulative additions of lead to agricultural soils be limited to a maximum of 800 kg ha^{-1} rather than the values shown in Table 5 (EPA-FDA-USDA, 1981).

For soils used for growth of animal feed only, neither annual nor cumulative cadmium application limits have been established, but soil pH must be 6.5 and a plan is needed to show that the crop will not directly enter the human diet.

A general statement of federal policy and guidance in relation to land application of municipal sewage sludge for the production of fruits and vegetables has been published (EPA-FDA-USDA, 1981). The three federal agencies agree that the use of high quality sludges, coupled with proper management procedures, should safeguard the consumer from contaminated crops and minimize any potential adverse effect on the environment.

In addition, some states have even more stringent guidelines concerning sludge application on the land. For instance, in 1988 the Pennsylvania Department of Environmental Resources (PDER) issued *Guidelines for Sewage Sludge Use for Land Reclamation* in Pennsylvania (Department of Environmental Resources 1988). These guidelines state that due to the high permeability of mine spoils and low retention of organic matter, sufficient nitrogen in excess of the crop requirement must be provided in order to establish growth. To provide sufficient nitrogen a maximum application rate of 134 metric tons ha^{-1} may be utilized for land reclamation. In addition, the application is further limited according to the trace metal content of the sludge and application rates may not exceed the limits given in Table 6.

The Pennsylvania state guidelines further require that the soil pH be adjusted to 6.0 during the first year of sludge application and maintained at 6.5 for two years following final sludge application. Liming is required to immobilize the

Table 6.   Pennsylvania Recommended Maximum Trace Element Loading Rates for Land Reclamation

Constituent	Maximum Loading Rate (kg ha⁻¹)	
	Land Reclamation	Land Reclamation for Farming
Cd	5.6	3.4
Cu	140.0	84.0
Cr	560.0	336.0
Pb	560.0	336.0
Hg	1.7	1.1
Ni	56.0	33.0
Zn	280.0	168.0

Source: Pennsylvania Department of Environmental Resources, 1988.

trace metals in order to reduce their availability for plant uptake and to prevent their leaching into groundwater.

Other requirements include the following:

1. Sludge is to be incorporated within 24 hr after application.
2. Sludge is not to be applied when the ground is saturated, snow covered, frozen, or during periods of rain.
3. Sludge is not to be applied within 30 m of streams, 90 m of water supplies, 8 m of bedrock outcrops, 15 m of property lines, or 90 m of occupied dwellings.
4. Sludge for revegetation of inactive mines or active coal refuse piles is not to be applied to slopes exceeding 15%.
5. Dairy cattle must not be allowed to graze land for at least two months after sludge application.

The potential for successful reclamation with municipal sludge is tremendous. Most of the highly beneficial properties of sludge as a soil amendment come from its high organic matter content. Although it has long been shown to increase productivity on good agricultural soils, sludge organic matter is extremely important where topsoil is inadequate in amount or quality, e.g., on sites that were previously in forest, and especially for areas mined and left barren for many years before the 1977 act, where no topsoil exists. It is on land like this that the benefits of sludge are most evident. The establishment as well as the continued maintenance of cover for the minimum required five years is difficult on acid, eroded, infertile, compacted, and stony spoils that have no buffering capacity for temperature extremes and are incapable of retaining sufficient moisture for plants during periods of water stress. Although organic matter is the single most important component in the improvement of soil physical properties, sludges also contain neutralizing compounds and fertilizer elements that improve spoil pH and fertility.

While the benefits of using sludge to reclaim land seem obvious, there is still some reluctance on the part of landowners, local citizens, and local government officials to accept its use for reclamation. Due to the nature of devastated lands, larger amounts of sludge are used than for farmlands, but usually only a single application is made which allows the vegetation to become self-sustaining. The greatest obstacle appears to be the lack of knowledge on the part of the general public about the possible impacts of sludge on soils, plants, groundwater, surface water, and animal and human health. These impacts must be known in order to make rational decisions concerning the benefits and risks.

There is a large amount of research and demonstration experience and information available on all types of land application projects, initiated to address public concerns. The increasing number of successful projects across the country clearly shows the value of sludge as a resource rather than disposing of it as a waste product.

Considerable detailed information is available on the planning, economics, social and legal problems, engineering, agricultural, ecological, and health-related aspects of reclaiming disturbed land with sewage sludge over a wide range of situations. Much of this information is summarized in the Process Design Manual for Land Application of Municipal Sludge (U.S. EPA, 1983) and in A Guide for Revegetation of Mined Land in Eastern United States Using Municipal Sludge (Sopper and Seaker, 1983) and in CRC Critical Reviews in Environmental Control (Sopper and Seaker, 1984a).

CHAPTER 2

Review of Land Reclamation Projects Using Municipal Sludge

OVERVIEW

During the past 20 years a considerable amount of research has been conducted on the feasibility of using municipal sludge for the revegetation of mine land. Some of the more significant projects are summarized in Table 7. Figure 2 shows the number of mine land reclamation projects using municipal sludge as an amendment by states cited in Table 7. An attempt has been made to review these publications and to summarize the results in terms of evaluating the effects of sludge applications on vegetation growth responses, vegetation quality, physical, chemical, and biological properties of the minesoil, soil percolate and groundwater quality, and animal nutrition and health.

EFFECTS ON VEGETATION
Growth Responses
Grass and Legume Species

The productivity and fertility of lands disturbed by mining activities has been substantially improved in most cases by sludge applications. Larger yield increases have been realized on sludge-amended mine land than on the same type of land amended with inorganic fertilizers.

In the Midwest, nine grass species were seeded on calcareous stripmine spoil with pH between 6.0 and 7.5. Tall fescue,[a] perennial ryegrass, and western wheatgrass showed the most rapid establishment and most vigorous growth where relatively high sludge loading rates (224, 448, and 896 Mg ha^{-1}) were applied

---

[a] A complete list of common names and scientific names of vegetation discussed in this chapter is given in Appendix Table A-1.

Table 7. Recent Land Reclamation Projects with Municipal Sludges

Type of Disturbed Land	State	Sludge Type[a]	Application Rates (Mg ha⁻¹)	Plant/Animal Studied	Parameters Tested[b]	Reference
Acid strip mine spoil	PA	Dig.-D +effluent	5–20 cm	Ryegrass, Hybrid poplar	WA	McCormick and Borden, 1973
Deep mine anthracite refuse	PA	Dig.-D	0,40–150	10 Tree spp., 5 Grass spp., 5 Legume spp.	GR PA SA WA	Kerr et al., 1979
Acid strip mine spoil and deep mine refuse	PA	Dig.-L + effluent	0–51 cm	8 Tree spp., 8 Grass spp., 8 Legume spp.	GR WA SA	Kardos et al., 1979
Acid strip mine spoil	PA VA	Dig.-L,D	198–730	Tall fescue, Red top, Ladino clover, Birdsfoot trefoil, Ryegrass	GR	Hill et al., 1979
Acid strip mine spoil	PA	Dig.-D,C	132	Tall fescue, Orchardgrass, Birdsfoot trefoil, Crownvetch, Alfalfa	GR PA SA WA	Murray et al., 1981
Acid strip mine spoil and deep mine refuse	PA	Dig.-D,L,C	0,7–202	Tall fescue, Orchardgrass, Birdsfoot trefoil, Crownvetch	GR PA SA WA	Sopper and Kerr, 1982; Sopper et al., 1981
Acid strip mine spoil and deep mine refuse	PA	Dig.,D,L,C	0,7–202	Tall fescue, Orchardgrass, Birdsfoot trefoil, Crownvetch	GR PA SA WA	Sopper and Seaker, 1982
Top-soiled strip mine spoil	PA	Dig.,D,C	202	Tall fescue, Orchardgrass, Birdsfoot trefoil, Crownvetch	GR PA SA WA	Murray et al., 1981

Site/material	State	Sludge	Rate	Vegetation	Code	Reference
Acid strip mine spoil; Top-soil strip mine spoil (49 sites)	PA	Dig.,D,C	134	Tall fescue, Orchardgrass, Birdsfoot trefoil, Ryegrass	GR, PA, SA, WA	Carello, 1990
Acid strip mine spoil and deep mine refuse	PA	Dig.-D,L,C	11–202	Birdsfoot trefoil, Tall fescue, Orchardgrass	PA	Seaker and Sopper, 1982
Deep mine anthracite refuse	PA	Dig.,D	0,80,108	Tall fescue, Orchardgrass, Birdsfoot trefoil, Crownvetch	GR, PA, SA, WA	Seaker and Sopper, 1983
Acid strip mine spoil	PA	Dig.,D,L	0,7,11,90,184	Tall fescue, Orchardgrass, Birdsfoot trefoil, Crownvetch	GR, PA, SA, WA	Seaker and Sopper, 1984
Acid strip mine spoil	PA	Dig.,D,C/C	134	Rabbit	PA,SA,AH	Dressler et al., 1986
Zinc smelter site	PA	Dig.,D	47	5 Grass species, 5 Legume species, 11 Tree species	GR, PA, SA	Sopper, 1989a; Sopper, 1989b; Sopper, 1989c; Sopper, 1987; Sopper and McMahon, 1987; Sopper and McMahon, 1988a; Sopper and McMahon, 1988b
Acid strip mine spoil	PA	Dig.,D,C/C	120–134	Microorganisms	SO,SA	Sopper and Seaker, 1987a; Sopper and Seaker, 1987b; Seaker and Sopper, 1988a

Table 7. Recent Land Reclamation Projects with Municipal Sludges (Continued)

Type of Disturbed Land	State	Sludge Type[a]	Application Rates (Mg ha^{-1})	Plant/Animal Studied	Parameters Tested[b]	Reference
Acid strip mine spoil	PA	Dig.,D,C/C	120–134	NA	SA	Seaker and Sopper, 1988b
Acid strip mine spoil	PA	Dig.,D,C/C	128	Microorganisms	SO,SA	Sopper and Seaker, 1988
Acid strip mine spoil	PA	Dig.,D,C/C	134	Vole	PA,SA,AH	Alberici et al., 1989
Coal mine spoil	CO	Dig.,D	0,14,28,55	8 Grass species	SA,PA GR	Topper and Sabey, 1986
Coal spoil and topsoil	CO	Dig.,D	0,40,80,120	Microorganism activity	SA	Voos and Sabey, 1987
Copper mine spoil	CO	Dig.,D	0,30,60	Fourwing saltbush Mountain big sagebrush	GR,PA,SA	Sabey et al., 1990
Copper mine Borrow pit Kaolin spoil Marginal land	TN GA	Dig.-D	0,34,69,275	Pine species Sweetgum	GR,SA	Berry, 1982
Acid strip mine spoil	IL	Dig.-L	0,31–121	Tall fescue Weeping lovegrass	GR WA	Boesch, 1974
Acid strip mine spoil	IL	Dig.-L	0,78,304	Tall fescue Weeping lovegrass	GR SA WA	Lejcher and Kunkle, 1974
Acid strip mine spoil	IL	Dig.-L	0,314,627	Tall fescue Alfalfa	GR PA	Stucky and Newman, 1977 Blessin and Garcia, 1979
Strip mine spoil	IL	Dig.-L	56	Corn	PA	Garcia, 1979
Strip mine spoil	IL	Dig.-L	0–997	Blackbirds	AH	Gaffney and Ellertson, 1979
Strip mine spoil	IL	Dig.-L	NI	NI	PO	Lue-Hing et al.,

Site	State	Application	Rate	Crop	Code	Reference
Strip mine spoil	IL	Dig.-L	NI	NI	SO	Sundberg et al., 1979
Strip mine spoil	IL	Dig.-L	25–128/yr	Corn	AH	Hinesly et al., 1979
Strip mine spoil	IL	Dig.-L	NI	Pheasant, swine Cattle	AH,PO SA,PA	Fitzgerald, 1979
Acid strip mine spoil	IL	Dig.-L	448–997	8 Herbaceous spp. 18 Tree spp.	WA	Jones and Cunningham, 1979
Acid strip mine spoil	IL	Dig.-L	448–997	7 Legumes 10 Grasses	GR PA SA	Stucky and Bauer, 1979
Acid strip mine spoil	IL	Dig.-L	448–997	12 Tree spp.	GR PA	Roth et al., 1979
Acid strip mine	IL	Dig.-L	448–997	8 Tree spp.		Svoboda et al., 1979
Calcareous strip mine spoil	IL	Dig.-L	0.8–85.8	Corn	GR PA SA WA	Peterson et al., 1979
Coal refuse	IL	D	225–900	Tall fescue Redtop Reed canarygrass	GR SA	Joost, et al., 1981
Acid strip mine spoil	KY	Dig.-D	0,34–269	Corn Soybeans	GR PA SA	Feuerbacher et al., 1980
Acid strip mine spoil	KY	(NI)	28–96	European alder Blacklocust Cottonwood Loblolly pine Northern red oak	GR PA SA	Schneider et al., 1981
Reconstructed prime farmland	KY	Dig.-D	0,22.4,448	Grain sorgham Corn	GR PA SA	Powel et al., 1988

Table 7.   Recent Land Reclamation Projects with Municipal Sludges (Continued)

Type of Disturbed Land	State	Sludge		Plant/Animal Studied	Parameters Tested[b]	Reference
		Type[a]	Application Rates (Mg ha⁻¹)			
Strip mine spoil	IL	Dig.-D	0,224,448,896	9 Grass species Corn Rye	GR PA SA	Hinesly et al., 1979; Hinesly et al., 1982; Hinesley and Redborg, 1984
Calcareous strip mine spoil	IL	Dig.L	0,6.4,12.7,25.4 mm Corn		PA	Hinesly et al., 1984
Acid strip mine spoil	IL	Dig.-L	448–997	5 Tree species	PA SA	Roth et al., 1982
Acid strip mine spoil	IL	Dig.-L	336–672	Annual rye Orchardgrass Tall fescue	WA	Urie et al., 1982
Calcareous strip mine spoil	IL	Dig.-L	0–453	Corn	GR PA WA	Peterson et al., 1982
Strip mine spoil	IL	Dig.-L	0,11–45	Grasses Legumes Cattle	GR PA SA AH PO	Fitzgerald, 1982
Calcareous strip mine spoil	IL	Dig.,L.	174	Earthworms	SO	Pietz et al., 1984
Acidic coal refuse	IL		225,450,900	Tall fescue Reed canarygrass Red top	SA,PA,GR	Joost et al., 1987
Acidic coal refuse	IL	Dig.,D,L	237,305	Bromegrass Tall fescue Alfalfa	SA,PA, WA,GR	Pietz et al., 1989 a,b,c

Material	State	Treatment	Rate	Vegetation		Reference
C and D Canal dredge material	DE	Dig.-D	168	KY bluegrass / Tall fescue / Red fescue / Weeping lovegrass	PA / SA / WA	U.S. EPA and Gannett-Fleming, 1990
C and D Canal dredge material	MD	Dig.-D	112	KY bluegrass / Tall fescue / Red fescue / Weeping lovegrass	PA / SA / WA	U.S. EPA and Gannett-Fleming 1990
C and D Canal dredge material	DE	Dig.,D	100	Tall fescue / Red fescue / KY bluegrass	PA / SA	Palazzo and Reynolds, 1991
Acid strip mine spoil	MD	Dig.,C	0,56–224	Tall fescue / Birdsfoot trefoil	GR,PA, SA	Griebel et al., 1979
Gravel spoils	MD	C	0,40–160	Corn / Beans	GR,PA, WA	Hornick, 1982
Acid strip mine spoil	OH	D	658	Forage	GR,PA	Sutton and Vimmerstedt, 1974
Acid strip mine spoil	OH	Dig.,-D	11–716	Tall fescue	GR,PA, SA,WA	Haghiri and Sutton, 1982
Degraded, Semiarid Grassland	NM	Dig.,D	0,22.5,45,90	Blue gamma / Galleta / Bottlebrush squirreltail	GR,PA, SA	Fresquez et al., 1990a; Fresquez et al., 1990b
Zn smelter surroundings	OK	Dig.-L +effluent	2.5–34 cm	10 Grass spp. / 1 Legume	GR / PA / SA	Franks et al., 1982
Lignite overburden	TX	Dig.,D	56	NA	SA,WA	Hornby et al., 1986
Lignite overburden	TX	Dig.,D	142,284	Bermudagrass	SA,PA	Cocke and Brown, 1987
Acid strip mine spoil	VA	C	0,159–412	Virginia pine / Tall fescue / Perennial ryegrass / Annual ryegrass	GR / SA	Scanlon et al., 1973
Acid strip mine spoil	VA	Dig.-D	Various (NI)	(NI)	SA	Younos and Smolen, 1981

Table 7.   Recent Land Reclamation Projects with Municipal Sludges (Continued)

Type of Disturbed Land	State	Sludge Type[a]	Sludge Application Rates (Mg ha⁻¹)	Plant/Animal Studied	Parameters Tested[b]	Reference
Abandoned Pyrite mine	VA	Dig.-D	82–260	Tall fescue Lespedeza Weeping lovegrass Wheat, rye, oats	GR PA SA WA	Hinkle, 1982
Sandstone and siltstone mine soil	VA	Dig.-D	22,56,112,224	Tall fescue	GR SA,PA	Roberts et al., 1988
None acid-forming overburden	VA	Dig.-D,C	112	Hay/pasture seed mix	SA WA	Daniels and Haering, 1990
Borrow pit	SC	Dig.-D	0,17,34,68	Tall fescue Sweetgum	GR PA	Kormanik and Schultz, 1985
Acid strip mine spoil	WV	D	0–224	Tall fescue	GR PA SA	Mathias et al., 1979
Acid strip mine spoil	WV	Dig.-D,C	Various (NI)	Blueberries	GR PA	Tunison et al., 1982
Overburden minesoil	WV	Dig.-D	0,22.4,44.8,78.4	Red clover Tall fescue Orchardgrass Birdsfoot trefoil	SA,PA GR	Skousen, 1988
Acid strip mine spoil	WV	Dig.,D	0,45,90	Alfalfa Lespedeza	SA,PA	Keefer et al., 1983
Iron ore tailings	WI	Dig.-D	42–85	5 Native prairie grasses 4 Prairie forbes Foxtail	GR	Morrison and Hardell, 1982
Taconite tailings	WI	Dig.-D	28–115	4 Grass-legume mixtures	GR	Cavey and Bowles, 1982
Colliery coal mine waste (Scotland) (Scotland)	U.K	NI	56,112	Herbage (NI)	GR SA	Pulford et al., 1988 Pulford, 1991

Opencast coal mine Ardoss site	U.K.	L	364,728 (m³ ha⁻¹)	Sitka spruce	GR	Bayes et al., 1990
Opencast coal mine Clydesdale site	U.K.	L,Dig.	94,195 (m³ ha⁻¹)	Sitka spruce	GR	Bayes et al., 1990
Opencast coal mine Clydesdale site	U.K.	Dig.-D	106,194 (m³ ha⁻¹)	Herbaceous sward	GR	Bayes et al., 1990
Opencast coal mine Clydesdale site	U.K.	Lig.-D	114,159 (m³ ha⁻¹)	Sitka spruce	GR	Arnot et al., 1990
Opencast coal mine site	Germany	C	4.7,9.4	N.I.	SA	Werner et al., 1990

[a] Dig. = digested, L = liquid, D = dewatered, C = composted, C/C = dewatered cake and compost mix, NI = no information.
[b] GR = growth responses, PA = plant tissue analysis, SA = soil analysis, WA = water analysis, SO = soil organisms, PO = pathogenic organisms, AH = animal health, NA = not applicable.

Figure 2.   Number of mine land reclamation projects using municipal sludge as an amendment cited in Table 7.

(Hinesly et al., 1982). In a greenhouse study, four different sludges were applied
to acidic spoils. Tall fescue yield increased with increasing rate of two of the
sludges, but high metal concentrations of the other two sludges inhibited fescue
growth (Haghiri and Sutton, 1982). In another greenhouse study Stucky and
Newman (1977) grew tall fescue and alfalfa for two years using 314 and 627 Mg
ha^{-1} of sludge in acidic spoils. The high application significantly increased yields,
especially for alfalfa. Where sludge was applied on Ohio mine spoil at 658 Mg
ha^{-1} without lime, forage yield the first growing season was greater than 4.5 Mg
ha^{-1} (Sutton and Vimmerstedt, 1974). On the Palzo stripmine site in Illinois, 16
seed mixtures were evaluated on plots receiving 426 to 997 Mg ha^{-1} of sludge
(Stucky and Bauer, 1979; Stucky et al., 1980). Perennial rye was a very successful
initial cover crop, while Bermuda, orchard, reed canarygrass, and tall fescue
grasses were recommended for permanent cover, along with the legumes red
clover, lespedezas, and birdsfoot trefoil. Boesch (1974) gives a vivid description
of the devastated Palzo site and the sequence of events that led to large-scale
reclamation with Chicago sludge. In 1970, demonstration plots were seeded with
tall fescue and weeping lovegrass, which did not germinate on control plots, but
produced a complete groundcover in two months where 271 Mg ha^{-1} of sludge was
applied. On coal refuse in Illinois, Joost and others (1981) observed a more than
adequate cover with tall fescue, redtop, and reed canarygrass two months after
seeding, even with high rates of high-metal sludge. Redtop was the most successful.

In the East, Kardos and others (1979) tested ten grass and ten legume species in boxes of bituminous mine spoil and anthracite refuse amended with liquid digested sludge. Sludge and effluent irrigation at 2.5 and 5.0 cm per week and totaling 59 to 147 cm per year detoxified the highly acidic materials and established a vegetation cover. The most successful species were weeping lovegrass, ladino cover, Iroquois alfalfa, sericea lespedeza, and red clover. Subsequently, Sopper and Kerr (1982) grew a lush cover of tall fescue, orchardgrass, birdsfoot trefoil, and crownvetch on several 4 ha demonstration sites in Pennsylvania. The sites were amended with several different types of municipal sludge applied at rates ranging from 7 to 202 Mg ha^{-1}. The rationale for the mixture is that the grasses provide quick cover while the legumes eventually take over to provide the permanent cover. Following the demonstrations, over 1500 ha have been successfully reclaimed using Philadelphia sludge, with phenomenal annual increases in dry matter production that surpass the Pennsylvania stripmine revegetation requirements.

Approximately 87% of these reclaimed mine sites have remained as wildlife habitat, while the remaining 13% have been returned to agricultural use, including hay and corn production, and grazing. Hunting of deer and small game occurs on most of the wildlife habitat areas.

Yields of tall fescue on sludge-amended mine spoil (112 and 224 Mg ha^{-1}) in West Virginia were over 11,000 kg ha^{-1}, an 818% increase over controls (Mathias et al., 1979). Both Mathias and others (1979) and Hinesly and others (1982) observed some yield decreases over three growing seasons when sludge was applied only once and the crop was harvested annually with no additional amendments. Where the vegetation was not harvested, however, the recycling of nutrients and buildup of organic matter resulted in annual yield increases (Sopper and Seaker, 1982). The slow-release fertilizing action of sludge provided a "residual effect" for forage species.

In Pennsylvania, dewatered and heat-dried municipal sludge from Scranton, PA was applied at rates of 0, 40, 75, and 150 Mg ha^{-1} to plots on a burned anthracite coal refuse bank (red gob) (Sopper, 1990a). After incorporation, ten species of grasses and legumes were broadcast seeded. The chemical analyses of the sludge is given in Table 8 and the amounts of nutrients and trace metals applied in the various sludge applications is given in Table 9. Average percent cover for all seeded grass and legume species are given in Table 10. Of the grass species, reed canarygrass, orchardgrass, and tall fescue had the best growth response. Penngift crownvetch and birdsfoot trefoil had the highest percent cover of the legume species after three growing seasons. Average dry matter production for all seeded grass and legume species are given in Table 11. Reed canarygrass, crownvetch, and birdsfoot trefoil had the highest productivity. By the third growing season blackwell switchgrass and ladino clover were completely gone and deertongue was barely surviving. Volunteer vegetation was greatly stimulated by the sludge treatments. By the end of the third growing season, all sludge-treated plots had over 95% total (seeded and volunteer species) cover (Table 12) and the

Table 8.   Chemical Analysis of the Sludge

Element	Mean Concentration Dry Weight Basis
pH	8.7
Total P	0.55%
Total N	2.16%
Dry solids	96%
	mg kg^{-1}
K	253
Ca	4215
Mg	3431
Na	548
Fe	2251
Al	2500
Mn	300
Cu	732
Ni	38
Zn	797
Co	22
Cr	207
Pb	686
Cd	1.3

Source: Sopper (1990a).

highest dry matter production (Table 13). After three growing seasons, percent areal cover and dry matter production were the highest on the 75 and 150 Mg ha^{-1} sludge-treated plots. Composted Washington, D.C., sludge at 112 Mg ha^{-1} produced forage cover in accordance with Maryland strip-mine laws (i.e., 80% cover and 10% legume species after two years). Rates lower than 112 Mg ha^{-1} were not as effective, but rates greater than 112 Mg ha^{-1} resulted in superior performance (Griebel et al., 1979). In Virginia, the same sludge grew a very

Table 9.   Amounts of Nutrients and Trace Metals Applied in the Sludge Applications

Constituent	Sludge Application Rate (Mg ha^{-1})			PDER Lifetime Maximum
	40	75	150	
	kg ha^{-1}			
N	871	1645	3193	—
P	222	419	813	—
K	10	19	37	—
Zn	32	61	118	224
Cu	30	56	108	112
Pb	28	52	101	112
Cr	8	15	30	112
Ni	1.5	2.9	5.6	22
Cr	0.05	0.09	0.19	3

Source: Sopper (1990a).

Table 10.   Average Percent Cover for All Seeded Grass and Legume Species After Three Growing Seasons

	Sludge Application Rate (Mg ha⁻¹)			
Species	0	40	75	150
Grasses				
Blackwall Switchgrass	0	0	0	0
Deertongue	38	5	7	5
Reed Canarygrass	48	92	95	97
Orchardgrass	18	88	90	93
Tall Fescue	22	82	92	90
Legumes				
Bristly Locust	48	15	12	2
Iroquois Alfalfa	5	5	2	2
Ladino Clover	0	0	0	0
Penngift Crownvetch	63	78	68	83
Birdsfoot Trefoil	63	47	40	35

Source: Sopper (1990a).

successful tall fescue and weeping lovegrass cover on two abandoned pyrite mines (Hinkle, 1982), but a less than adequate cover on another site (Hill et al., 1979) due to low pH and drought conditions. The same author, however, observed threefold yield increases over plots amended with a standard inorganic fertilizer when Pennsylvania coal spoils were amended with Williamsport sludge at 730 Mg ha⁻¹.

In Virginia, 159 and 412 Mg ha⁻¹ of composted sludge without limestone amendments established a good cover of tall fescue and perennial and annual ryegrass on stripmine spoil, along with volunteer native species. By the fourth growing season, results were excellent whereas those on plots amended with inorganic fertilizer were poor. In another study, tall fescue, annual ryegrass, and

Table 11.   Average Dry Matter Production of the Seeded Grass and Legume Species After Three Growing Seasons

	Sludge Application Rate (Mg ha⁻¹)			
Species	0	40	75	150
	kg ha⁻¹			
Grasses				
Blackwell Switchgrass	0	0	0	0
Deertongue	535	222	387	420
Reed Canarygrass	1701	6519	6408	9120
Orchardgrass	922	3115	4578	6157
Tall Fescue	825	3760	5077	7664
Legumes				
Bristly Loucst	1396	549	1055	355
Iroquios Alfalfa	362	1468	1019	919
Ladino Clover	0	0	0	0
Penngift Crownvetch	1937	3283	3326	5070
Birdsfoot Trefoil	1586	2712	3731	5518

Source: Sopper (1990a).

Table 12.   Average Percent Cover for All Seeded Grasses and Legumes and for All Herbaceous Vegetation on the Plots

Sludge Application (Mg ha⁻¹)	First Year	Second Year	Third Year
Seeded Grasses and Legumes (%)			
0	36	41	30
40	31	48	41
75	28	49	41
150	18	44	38
Seeded and Volunteer (%)			
0	41	72	77
40	39	85	96
75	35	89	98
150	22	89	100

Source: Sopper (1990a).

sericea lespedeza produced a significantly denser cover with sludge than with inorganic fertilizer. Again, volunteer species were more abundant on the sludged plots. At 58 Mg ha⁻¹ the sludge was more effective than at 31 Mg ha⁻¹, which points out the need to supply sufficient plant nutrients (Scanlon et al., 1973).

In West Virginia, Skousen (1988) investigated the effects of applications of Morgantown sludge (Morganite) applied on stripmine spoil at rates of 22.4, 44.8, and 78.4 Mg ha⁻¹ on vegetative cover and biomass. Total vegetative cover was significantly higher on sludge-treated plots than on control plots and grass biomass increased with increasing sludge rates. Total cover percent were 60, 77, 84,

Table 13.   Average Dry Matter Production of All Herbaceous Vegetation

Sludge Application Mg ha⁻¹	First Year	Second Year	Third Year	Mean
Seeded Grasses and Legumes (kg ha⁻¹)				
0	379	1016	1567	987
40	1189	2162	2327	1893
75	1336	2558	2913	2269
150	402	3522	3683	2536
Seeded and Volunteer (kg ha⁻¹)				
0	407	2388	3338	2044
40	1471	5124	5539	4045
75	1647	6108	8044	5266
150	698	8646	7359	5568

Source: Sopper (1990a).

and 95 and grass biomass was 72, 126, 150, and 223 g per 0.25 square meter quadrat, respectively, for the four application rates. Legume and weed biomass decreased as sludge rates increased because increased nutrient levels in the minesoil caused increased grass growth and competition. Foliar analyses were not conducted but no visual symptoms of stress or reduced growth of plants in the sludge-treated plots were observed.

In the United Kingdom, the potential for using sewage sludge for land reclamation is large. A 1974 survey showed that there were about 43,000 ha of derelict land of which 33,000 ha justified restoration (Department of Environment, 1975). It is estimated that only approximately 7% of the UK annual sludge production (60,000 tonnes dry solids) was used in land reclamation in 1975 (Department of the Environment/National Water Council, 1981). Additional information on sludge utilization programs in the United Kingdom can be found in a recent book published by Pergamon Press and edited by Hall (1991).

Coker et al. (1982) reported on a project to demonstrate the beneficial effects of using sludge for land reclamation. The site used was a former gravel pit at Smallford, Hertfordshire that had been filled with urban refuse and capped with a 1-m layer of subsoil. Air-dried digested sludge from the Thames Water Authority was applied to plots at 35, 70, 140, and 280 Mg ha^{-1}. In addition, lime-treated undigested (raw) press cake was also applied to plots at 20 and 160 Mg ha^{-1}. After incorporation, plots were seeded with a mixture of fodder oats and perennial ryegrass. Forage yields the first year (1977) after sludge application were 2.94, 4.19, 6.59, 7.22, and 7.68 for the 0, 35, 70, 140, and 280 Mg ha^{-1} sludge application rates, respectively. Even after four years, forage yields were 2.21, 2.55, 2.17, 3.17, and 4.18 for the five sludge application rates, respectively. Similarly, the lime-treated undigested fresh sludge cake produced forage yields the first year (1977) of 1.77, 3.15, and 8.32 for the sludge application rates of 0, 20, and 120 Mg ha^{-1}, respectively. In the fourth year, forage yields were 2.13, 2.35, and 4.25 Mg ha^{-1} for the three application rates, respectively.

Only sludge application at 280 Mg ha^{-1} significantly increased the N content of the forage the first year. Amounts of trace metals applied in the highest sludge application (280 Mg ha^{-1}) were 19.6 kg Cd ha^{-1}, 291.2 kg Cu ha^{-1}, 61.6 kg Ni ha^{-1}, and 509.6 kg Zn ha^{-1}. Cadmium concentrations in forage the first year ranged from 0.07 to 1.02 mg kg^{-1} in comparison to 0.07 to 0.26 mg kg^{-1} in control forage. Forage Cu concentrations were 8.20 mg kg^{-1} on the plot where 291.2 kg Cu ha^{-1} was applied. Nickel concentrations in the forage ranged from 5.6 to 4.00 mg kg^{-1} on the high sludge application plot. Forage Zn concentration increased to 69.8 mg kg^{-1} on the high sludge application plot in comparison to 35.6 mg kg^{-1} on the control plot. A comparison of forage trace metal concentrations resulting from the highest sludge application rate (280 Mg ha^{-1}) with plant and animal deficiency thresholds, and phytotoxic and zootoxic thresholds showed that at no stage did concentrations of metals in forage give cause for concern, indicating that single, heavy applications of sludge of the order of 100 Mg ha^{-1} needed to restore soil fertility will not lead to metal contamination problems.

These comparisons were as follows:

Thresholds and Sludge Application Rates	Trace Metal mg kg^{-1}			
	Cd	Cu	Ni	Zn
Lower critical levels				
deficiency threshold (plants)	—	4	—	20
Deficiency threshold (animals)	—	10	—	50
Upper critical levels				
Phytotoxic threshold	8	20	11	200
Zootoxic threshold	3	30	50	500
Highest observed levels				
280 Mg ha^{-1}	1.02	8.2	5.6	70
0 Mg ha^{-1}	0.26	7.3	3.7	42

Source: Adapted from Coker et al. (1982).

Significant increases in soil organic matter and decreases in bulk density following sludge applications were also reported. Soil organic matter was increased by 1.1, 2.3, 4.7, and 9.3% by the addition of sludge at 35, 70, 140, and 280 Mg ha^{-1}. Bulk density was 1.79, 1.75, 1.70, and 1.46 for the four sludge applications, respectively in comparison to 1.78 for the control plot.

Metcalfe (1984) and Metcalfe and Lavin (1991) have reported on the use of consolidated sewage as soil substitute in colliery spoil reclamation at Mill Bank. Sewage sludge was applied on colliery spoil and mixed to create a seed bed composed of a 1:1 (wet volume) mixture of spoil and sludge, equivalent to a 2:1 mixture by dry weight. Sludge used was lime conditioned press cake applied at 656 Mg ha^{-1}. The area was seeded with the agricultural grass/clover mixture (Nickersons Red Circle) at 20g m^2. Amounts of metals applied in the sludge were 695 kg Cr ha^{-1}, 25 kg Ni ha^{-1}, 610 kg Cu ha^{-1}, 485 g Zn ha^{-1}, 1.1 kg Cd ha^{-1}, and 335 kg Pb ha^{-1}. Sludge application did not significantly affect trace metal concentrations in forage over a four-year period of monitoring. Concentrations of Zn were less than 5, Cu less than 2, Pb less than 18, Ni less than 3, Cr less than 1.5, and Cd less than 1.0 mg kg^{-1}. All of the element concentrations were less than critical levels cited by Davis and Beckett (1978) for ryegrass. Critical concentrations (mg kg^{-1}) cited for yield reductions were 10 for Cr, 14 for Ni, 21 for Cu, 221 for Zn, 10 for Cd, and 35 for Pb.

At another site, East Bierly, lagoon dried sludge was applied at 692 Mg ha^{-1} (Metcalfe and Lavin, 1991). Sludge was cultivated into the spoil to a depth of 200 mm. It was then seeded with the amenity grass mixture, Mommersteeg 22, at 20g m^2. Amounts of metals applied in the sludge were 1211 kg Cr ha^{-1}, 50 kg Ni ha^{-1}, 253 kg Cu ha^{-1}, 1035 kg Zn ha^{-1}, 0.7 kg Cd ha^{-1}, and 372 kg Pb ha^{-1}. Forage concentrations of Ni, Zn, Pb, Cd, and Cr increased slightly but were still well below the upper critical concentrations for yield suppression cited by Davis and Beckett (1978). The only element that exceeded the upper critical concentration was Cu at 22.8 and 33.4 mg kg^{-1} over the five-year period of monitoring. However, forage samples taken from surrounding fields were found to have Cu concentrations ranging from 22.0 to 24.5 mg kg^{-1}, despite that these grasses were growing on normal pasture soils uncontaminated by sewage sludge. The general performance of the two sites was very good with even swards developing within two months of seeding and no fertilizer applications have been required since establishment to maintain productivity. Metcalfe and Lavin (1991) concluded that consolidated sewage sludges can form substitutes for topsoil in the reclamation of colliery wastes. The advantages to the Local Authority include an inexpensive alternative to soil and cost savings in both lime and fertilizer, while for the supplying Water Authority it is a convenient outlet for large quantities of sludge.

Two studies on iron ore tailings revegetation in Wisconsin found sludge to be far superior to inorganic fertilizers. Some of the successful species included sideoats grama grass, Canada wild rye, foxtail grass (Morrison and Hardell, 1982), smooth brome grass, alsike clover, and alfalfa (Cavey and Bowles, 1982). Soil contaminated in the vicinity of a zinc smelter in Oklahoma was successfully revegetated with sludge reinforced with urea-N, whereas inorganic fertilizer was ineffective. Switchgrass and kleingrass showed excellent response over three growing seasons (Franks et al., 1982).

Topper and Sabey (1986) reported that sewage sludge applications (0, 14, 28, 55, and 83 Mg ha^{-1}) on coal mine spoil in Colorado significantly increased the aboveground biomass and percent canopy cover of a pasture grass mixture over a control for two seasons of growth. Total production on the sludge treatments increased linearly up to 4300 kg ha^{-1} at the highest rate. The lowest sludge application rate (14 Mg ha^{-1}) produced 2400 kg ha^{-1} of aboveground biomass in comparison to 300 kg ha^{-1} on the control plots.

Fresquez et al. (1990a, 1990b) reported on a study where dried, anaerobically digested sludge from Albuquerque was applied at 0, 22.5, 45, and 90 Mg ha^{-1} to a degraded semiarid grassland site to determine the effects on vegetative yields and quality over four growing seasons. Total plant foliar cover and herbaceous plant yields increased significantly with sludge amendment, particularly at the 45 Mg ha^{-1} rate, as compared to controls. Blue gamma yields increased two- to threefold in amended plots over control plots. The forage quality of blue gamma, galleta, and bottlebrush squirreltail improved significantly as the levels of tissue N, P, K, and crude protein increased linearly with sludge application. Levels of Cd and Pb in all plant tissues did not increase significantly as a result of the sludge applications. Micronutrients, particularly Cu, Mn, and Zn, increased linearly in all the plants sampled after sludge application. Plant density, species richness, and diversity all decreased with increasing sludge rates. All macronutrients increased to acceptable levels in deficient plant tissues.

In Illinois, successful establishment of three forage grasses (reed canarygrass, tall fescue, and redtop) was achieved with applications of dried sludge at rates of 225, 450, and 900 Mg ha^{-1} on coal refuse. Most treatments maintained greater than 80% cover over the four-year study period. Herbage yield was not significantly different among amended treatments for any of the three grasses (Joost, et al., 1987).

Roberts et al. (1988) studied the effects of sewage sludge applications on a 2:1 sandstone-siltstone spoil mixture to facilitate establishment of tall fescue in Virginia. Aerobically digested sludge was applied on plots at rates of 22, 56, 112, and 124 Mg ha^{-1}. Tall fescue production generally increased with increasing sludge application rates. Treatments amended with >56 Mg ha^{-1} sludge supported about twice as much fescue growth as other treatments for all five years of the study. First-year standing biomass was 3.9, 7.7, 9.3, and 10.7 Mg ha^{-1} for the four sludge application rates, respectively in comparison to 5.8 Mg ha^{-1} on the control plot.

In Illinois, anaerobically digested sludge, lime, and gypsum and various mixtures of these amendments were applied to plots on coal refuse material and seeded with alfalfa, bromegrass, and tall fescue (Pietz et al., 1989b). Plant yields were measured for three years (1978–1980). Plant cover and dry matter yields increased each year on all sludge-treated plots. In 1980, the sludge (542 Mg ha^{-1}) and lime treatment had the highest percent cover (89%) and highest dry matter yield (6 Mg ha^{-1}). Nurse crops including wheat, rye, and oats (Hinkle, 1982) and barley (Cavey and Bowles, 1982) have been shown to promote vegetative cover significantly.

Haering and Powell (1991) have reported on a project in Virginia using Philadelphia sludge (mine mix) for reclamation of surface mined area. Mine mix is a 50:50 mixture of composted sewage sludge (aerated static pile composting method) and stabilized anaerobic sludge cake. The application rates were 92, 184, 368, and 552 Mg ha^{-1} of mine mix equivalent to 25, 50, 100, and 150 Mg ha^{-1} of sludge cake. The site was seeded with a mixture of grass and legume species (unspecified). Forage yields increased with sludge application rates. Forage yields were 1.54 Mg ha^{-1} for the control area (no treatment) to 4.79 for a fertilized area and to 4.23, 6.36, 5.68, and 7.67 Mg ha^{-1} for the 92, 184, 368, and 552 Mg ha^{-1} sludge treatments, respectively. Sludge applications also significantly increased soil pH, and concentrations of N, P, K, Ca, Mg, and Zn in the surface soil. Sludge applications only affected NO_3-N concentrations in groundwater at one monitoring well located in the center of the sludge treated area. The highest concentration of NO_3-N was 11 mg l^{-1} which slightly exceeded the maximum for drinking water. That value was recorded only for one month following 18 months of sampling after sludge application.

Sopper (1989a) has reported excellent success on using mixtures of municipal sewage sludge and fly ash on a highly contaminated site at a zinc smelter located in Palmerton, PA. It was quite obvious that attempting to plant or direct seed vegetation on the existing soil surface would be futile since natural revegetation had not occurred over the past 50 years. It is not possible for grass, legume, and tree species to germinate and survive when seeded directly on the highly contaminated soil. Thus, it was concluded that some type of growing medium had to be placed on top of the contaminated soil. Some possibilities were top soil, peat-vermiculite potting mixtures, and waste products. Economics eliminated the first two possibilities. So, attention was focused on local waste products. It was decided to investigate sludge and fly ash because of the potentially large volumes that might be needed. The hypothesis evaluated was to determine if vegetation germination, survival, and growth could be facilitated by using various mixtures of sludge and fly ash as a growing medium on top of the contaminated soil which

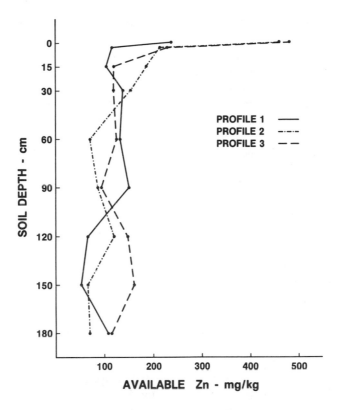

Figure 3.   Available Zn in the soil profile on the Palmerton zinc smelter site.

would provide a nutrient support system for the vegetation until some part of the root system penetrated the highly contaminated surface soil (15 cm) to reach the uncontaminated subsoil. The assumption was that vegetation survival would be greatly enhanced once the roots reached the normal uncontaminated subsoil. Soil samples taken in huge erosion gullies to a depth of 180 cm indicated that the depth of adverse contamination for Zn ($\pm$ 200 mg kg^{-1}) and Cd ($\pm$ 5 mg kg^{-1}) was about 15 cm (Figures 3 and 4).

Since the sludge could not be incorporated, it did not seem feasible to apply sludge alone. Its poor physical characteristics (wetting and drying), when applied on the soil surface without incorporation, would not be conducive to seed germination. Thus, it was decided to use the fly ash in a mixture with sludge to make a more friable product and more permeable to precipitation. Based upon analyses of the sludge (S) and fly ash (FA) it was decided to evaluate three mixtures of sludge and fly ash. The mixture ratios (on volume basis) were 1S:1FA, 2S:1FA, and 3S:1FA.

The amendments, along with 22 Mg ha^{-1} of lime and 90 kg ha^{-1} of potassium, were applied to nine 0.4 ha plots on the midslope area of Blue Mountain with an Estes "Aerospreader" truck. Typical chemical analyses of the sludge and fly ash

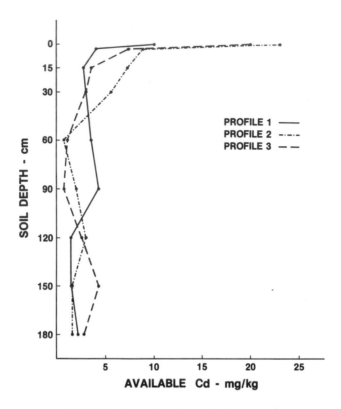

Figure 4.   Available Cd in the soil profile on the Palmerton zinc smelter site.

are given in Table 14 and the amounts of constituents applied in each mixture are given in Table 15, along with PDER maximum limits. The amounts of trace metals applied were well below the allowable maximum amounts recommended by PDER. Each 0.4-ha field plot was subdivided into five subplots for hydroseeding of five different seed mixtures (see Table 16; Sopper, 1989a). Fiber mulch was applied at 2240 kg ha⁻¹ with a hydroseeder.

Vegetation percent cover and dry matter production were measured at the end of each growing season (1986 and 1987). Average percent cover for each seeding mixture is given in Table 16. Of the five original seed mixtures, the highest percentage vegetation cover (81%) was obtained with the birdsfoot trefoil and tall fescue mixture and the pennfine perennial ryegrass and lathco flatpea mixture. The highest percentage covers were observed on the 1S:1FA and 2S:1FA. The mixture of orchardgrass, tall fescue, and crownvetch, selected on the basis of its successful use on sludge reclaimed mine land, overseeded on the cave-in-rock switchgrass subplots produced the highest percentage cover of all mixtures evaluated. In 1986, this seed mixture had an average cover of 90% and ranged from 82 to 95% among the three sludge-fly ash amendments. In 1987, average cover increased to 93%. The average percentage covers for all seed mixtures were higher in 1987 than in 1986.

Table 14.   Typical Chemical Analyses of the Sludge and Fly Ash

Constituent	Sludge Concentration	Fly Ash Concentration
pH	8.3	8.9
	%	mg kg^{-1}
Total P	1.06	70
Total N	3.59	—
NH$_4$-N	0.45	—
Org-N	3.14	—
Ca	5.10	5240
Mg	0.35	72
Na	0.12	88
K	0.23	211
Al	1.99	23
Fe	2.94	25
B	—	12
	mg kg^{-1}	
Mn	839	2.5
Zn	1763	15
Cu	747	3.8
Pb	154	1.2
Cr	379	43
Ni	94	0.3
Cd	9	0.2
Hg	1.4	—
Solids (%)	23	—
Soluble salts	—	227

Source: Sopper (1989a).

No measurements of dry matter production were made in 1986 because deer browsing on the plots was so heavy it made taking yield samples virtually impossible. However, in July 1987, yield samples were collected. The highest dry matter production (5761 kg ha^{-1}) was produced by the orchardgrass, tall fescue, and crownvetch mixture (Table 17). Three of the highest average yields were measured on the 2S:1FA amended plots.

More recently, Sopper (1991b) reported on herbaceous vegetation growth responses on a steep coal refuse bank amended with a mixture of sludge and fly ash. Excelsior matting was installed on plots on the refuse bank (slope 75%). Lime at 2800 kg ha^{-1} and the sludge-fly ash amendment were applied to three plots with an Estes "Aerospreader" truck. The sludge-fly ash mixture consisted of two parts of sludge to one part of fly ash (2S:1FA). Three adjacent control plots received lime and a 10-10-10 fertilizer applied at 2240 kg ha^{-1}. All plots were mulched with straw at 2240 kg ha^{-1} and tacked down with Cellin fiber mulch at 1350 kg ha^{-1}. All plots were hydroseeded with a mixture of grasses and legumes consisting of K-31 tall fescue, orchardgrass, birdsfoot trefoil, and crownvetch.

Table 15.   Sludge–Fly Ash Mixture Application Rates and Amounts of Chemical Constituents Applied per Hectare

Constituent	PDER Maximum Limits	Amount Applied (kg ha⁻¹)		
		1S:1FA[a]	2S:1FA[b]	3S:1FA[c]
Total P		1214	1120	991
Total N		2455	3030	2735
NH_4-N		282	297	314
Org-N		2172	2733	2421
Ca		5419	5466	3563
Mg		620	626	326
Na		197	148	101
K		1157	560	401
Al		9878	4791	3700
Fe		10131	6635	3939
Mn		43	464	26
Zn	224	100	109	59
Cu	112	68	83	53
Pb	112	26	19	16
Cr	112	46	39	34
Ni	22	22	18	10
Cd	3	0.9	0.9	0.8
Hg		0.1	0.1	0.1

Source: Sopper (1989a).

[a] 47 dtS = 235 dtFA; 235 wtS = 235 dtFA.
[b] 47 dtS = 117.6 dtFA; 235 wtS = 117.6 dtFA.
[c] 47 dtS = 78 dtFA; 235 wtS = 78 dtFA.

Table 16.   Average Percentage Vegetation Cover

Seed Mixture	Subplot	Amendment Mixture			1986	1987
		1S:1FA	2S:1FA	3S:1FA		
Blackwell switchgrass[a]	1	98[d]	19	—	40	59
Cave-in-rock switchgrass[b]	1	—	—	2	<1	2
Niagara big bluestem	2	53	39	32	21	41
Birdsfoot trefoil and tall fescue	3	95	98	49	74	81
Pennfine perennial ryegrass and Lathco flatpea	4	73	88	81	46	81
Oahe intermediate wheatgrass	5	23	45	12	16	27
Orchardgrass, tall fescue and crownvetch[c]	1	91	90	97	90	93

Source: Sopper (1989a).

[a] Seeded on Field Plots 2, 4, and 6.
[b] Seeded on Field Plots 3, 5, 7 and 9.
[c] Overseeded on Subplot 1 on Field Plots 3, 5, and 9 in July 1986.
[d] Mostly invading volunteer vegetation.

Table 17.   Average Dry Matter Production (kg ha⁻¹) for Each Seed Mixture in 1987

Seed mixture	Subplot	Amendment Mixture			Average
		1S:1FA	2S:1FA	3S:1FA	
Blackwell switchgrass	1	4410	2432	—	3421
Cave-in-rock switchgrass	1	—	—	83	83
Niagara big bluestem	2	1932	1648	1908	1829
Birdsfoot trefoil and tall fescue	3	5260	6201	3166	4876
Pennfine perennial ryegrass and Lathco flatpea	4	3076	6469	3931	4492
Oahe intermediate wheatgrass	5	5144	2883	995	3007
Orchardgrass, tall fescue and crownvetch[a]	1	4982	7454	4848	5761

Source: Sopper (1989a).

[a] Overseeded on Subplot 1 on Field Plots 3, 5, and 9

Table 18.   Chemical Analyses of the Sludge and Fly Ash

Constituent	Sludge Concentration (%)	Fly Ash Concentration[a] (mg kg⁻¹)
Total P	2.08	—
Bray P	—	20
Total N	4.88	—
NH₄-N	0.79	—
Org-N	4.09	—
Ca	5.76	6000
Mg	0.45	72
Na	0.11	64
K	0.12	222
Al	1.22	105
Fe	1.50	43
B	—	2.7
	mg kg⁻¹	
Mn	216	2.1
Zn	1243	0.9
Cu	1119	1.3
Pb	187	1.1
Cr	133	78
Ni	61	0.6
Cd	18	0.03
Hg	9.1	—
Solids (%)	17	—
pH	8.1	9.4

Source: Sopper (1991b).

[a] Concentrations are for available metals.

Table 19.   Chemical Analyses of the Sludge–Fly Ash Mixture and the Amounts of
Chemical Constituents Applied

Constituent	Concentration	Amount Applied (kg ha⁻¹)	PDER Maximum (kg ha⁻¹)
pH	8.4		
	%		
Solids	43		
Total P	0.64	1505	
Total N	1.27	2987	
NH_4-N	0.23	541	
Org-N	1.04	2446	
Ca	2.40	5646	
Mg	0.24	564	
Na	0.06	141	
K	0.35	823	
Al	2.47	5809	
Fe	5.61	13195	
	mg kg⁻¹		
Mn	314	74	
Zn	443	104	280
Cu	357	84	140
Pb	88	20	560
Cr	75	18	560
Ni	73	17	56
Cd	5.3	1.2	5.6
Hg	1.2	0.3	1.7

Source: Sopper (1991b).

The chemical analyses of the sludge and fly ash used on the project are given in Table 18. The sludge was high in nutrients (N and P) and had low trace metal concentrations. The fly ash had a high concentration of Ca and low concentrations of trace metals and boron. Soluble salts were high but were not expected to be a problem since the concentration will be diluted when mixed with the sludge. The amounts of chemical constituents applied in the sludge-fly ash mixture are given in Table 19. The sludge application rate was equivalent to 65 dry Mg ha⁻¹. The amounts of trace metals applied in the mixture were well below the allowable maximum amounts recommended by the Pennsylvania Department of Environmental Resources (PDER) for mine land reclamation (Table 19.)

Table 20.   Average Percent Cover and Dry Matter Production

	Control 1989	Sludge–Fly Ash 1989	Control 1990	Sludge–Fly Ash 1990
Percent cover (%)	6.9	57.3[a]	10.6	62.3[a]
Dry matter production (kg ha⁻¹)	799	5019[a]	824	6029[a]

Source: Sopper (1991b).

[a] Significant effect at $P < 0.001$.

Average percent cover and dry matter production are given in Table 20. Both percent cover and dry matter production were significantly higher in both years on the sludge-fly ash amended plots.

In Scotland, Pulford et al. (1988) and Pulford (1991) assessed the effects of various organic treatments on forage yields and nitrogen, phosphorus, and potassium concentrations in forage. Chicken manure, sewage sludge, a seaweed-based soil conditioner, and peat were applied to limed and unlimed plots at rates of 56 and 112 Mg ha^{-1}. All plots were also treated with 300 kg ha^{-1} of a 15-10-10 NPK fertilizer. They reported that the applications of chicken manure and sewage sludge resulted in much higher forage yield initially, but these declined so that by the seventh year there was no difference between any treatment and the controls. Forage yields, expressed as a percentage of the control plots, were 722% for the chicken manure, 226% for the sewage sludge, 34% for the soil conditioner, and 62% for the peat the first growing season. In the seventh year, forage yields were 85, 116, 150, and 99% of the control plot yields, respectively.

In Germany, Werner et al. (1990) reported on the use of sewage sludge to reclaim open cast mining areas in the "Rhenish brown-coal district." Sewage sludge and composted garbage and sewage sludge was applied on sites at 4.7 and 9.4 Mg ha^{-1} per year starting in 1969. Both N and C content of the soil were significantly increased in the 0 to 30-cm depth by the 9.4 Mg ha^{-1} application of composted garbage and sewage sludge after six years. The NO_3-N content (kg ha^{-1}) of the soil in the 0 to 90-cm depth in spring was always the highest in the treatment where sewage sludge alone was applied.

In general, studies show that good plant cover can be established on many types of disturbed land using municipal sludge, which is superior to inorganic fertilizer in such situations. Of course, plant performance varies considerably with species, and an appropriate seed mixture should be chosen carefully. Planting date is also crucial. On the Palzo site, winter rye seeded one month late resulted in a significant decrease in forage yield measured the following spring (Stucky and Bauer, 1979). Table 21 lists some successful plant species used in various sludge reclamation projects. Differences in soil and climate influence mixtures chosen for revegetation. For example, tall fescue is an excellent choice on highly acidic stripmine spoil in Pennsylvania (Sopper and Kerr, 1982), but that species failed completely on sludge-amended soils contaminated by a nearby zinc smelter in Oklahoma (Franks et al., 1982). There, kleingrass and switchgrass were most successful.

Field Crops

Several studies have looked at yields of corn, beans, and even blueberries grown on sludge-amended disturbed land. On sand and gravel spoils with low pH and a heterogeneous profile, sweet corn, field corn, and bush bean biomass were significantly increased by applications of Washington, DC, sludge. In general, bean yields were increased by 40 to 80 Mg ha^{-1} applications, but corn ear yields were not (Hornick, 1982). In Illinois, the yields of corn, soybeans, small grains, and forages over two to seven years were highly variable where Chicago sludge at rates up to 453 Mg ha^{-1} were applied to stripmine soils. Yields ranged from poor

Table 21.   Some Successful Plant Species and Species Mixtures Used in Various Sludge-Reclamation Projects

State	Species	Seeding Rate (kg ha⁻¹)	Reference
CO	Slender wheatgrass[a]	5.1	Topper and Sabey, 1986
	Intermediate wheatgrass[a]	4.8	
	Pubescent wheatgrass[a]	4.6	
	Crested wheatgrass[a]	3.8	
	Smooth brome[a]	4.6	
	Meadow brome[a]	2.6	
	Timothy[a]	1.5	
	Orchardgrass[a]	1.4	
IL	Tall fescue	22	Lejcher and Kunkle, 1973
	Weeping lovegrass	8	
IL	K-31 tall fescue	22	Boesch, 1974
	Weeping lovegrass	7.8	
IL	Common bermudagrass[a]	11	Stucky and Bauer, 1979
	Sericea lespedeza[a]	28	
	Kobe lespedeza[a]	11	
	Perennial rye grass[a]	22	
	Potomac orchardgrass[b]	17	
	Sericea lespedeza[b]	22	
	Kobe lespedeza[b]	11	
	Potomac orchardgrass[c]	22	
	Penngift crownvetch[c]	17	
IL	Tall fescue	25	Hinesly et al., 1982
	Perennial ryegrass	25	
	Western wheatgrass	25	
IL	Reed canarygrass	34	Joost et al., 1987
	Tall fescue	46	
	Redtop	17	
IL	Alfalfa[a]	22.9	Pietz et al., 1989b
	Bromegrass[a]	9.5	
	Tall fescue[a]	9.1	
MD	Tall fescue	40	Griebel et al., 1979
	Birdsfoot trefoil	10	
OH	Fall, balbo rye	9.6 (bu ha⁻¹)	Sutton and
	Spr., K-31 tall fescue[a]	11	Vimmerstedt, 1974
	Korean lespedeza[a]	3.4	
	Sweet clover[a]	3.5	
	Orchardgrass[a]	3.3	
OK	Switchgrass	154	Franks et al., 1982
	Kleingrass	154	
PA	Reed canarygrass	224	Kerr et al., 1979
	Tall fescue	224	
	Orchardgrass	224	
	Birdsfoot trefoil	224	
	Crownvetch	224	
	Deertongue	224	
	Switchgrass	224	
	Alfalfa	224	
	Ladino clover	224	
PA	Ky-31 tall fescue[a]	39	Hill et al., 1979
	Birdsfoot trefoil[a]	8	
	Rye grass[a]	6	
PA	Tall fescue[a]	22	Sopper and Kerr, 1982
	Orchardgrass[a]	22	
	Birdsfoot trefoil[a]	11	

Table 21.   Some Successful Plant Species and Species Mixtures Used in Various Sludge-Reclamation Projects (Continued)

State	Species	Seeding Rate (kg ha⁻¹)	Reference
	Crownvetch[a]	11	
PA	Blackwell switchgrass	17	Sopper, 1989a
	Niagara big bluestem	34	
	Birdsfoot trefoil	22[a]	
	K-31 tall fescue	45[a]	
	Perennial ryegrass	22[b]	
	Lathco flatpea	67[b]	
	Oahe intermediate Wheat grass	34	
	Tall fescue	22[c]	
	Orchardgrass	22[c]	
	Crownvetch	17[c]	
PA	K-31 tall fescue	22[a]	Sopper, 1991b
	Orchardgrass	22[a]	
	Birdsfoot trefoil	11[a]	
	Crownvetch	11[a]	
VA	Tall fescue[a]	8.4	Scanlon et al., 1973
	Perennial ryegrass[a]	8.4	
	Annual ryegrass[a]	8.4	
	Tall fescue[b]	22	
	Perennial rye[b]	22	
	Sericea lespedeza[b]	22	
	Black locust[b]	0.8	
VA	K-31 tall fescue[a]	67	Hill et al., 1979
	Redtop[a]	5.6	
	Ladino clover[a]	5.6	
VA	Tall fescue	67.3	Hinkle, 1982
	Weeping lovegrass	22	
	Korean lespedeza	11.2	
VA	Ky-31 tall fescue	80	Roberts et al., 1988
WI	Canada bluegrass[a]	11	Cavey and Bowles, 1982
	Red clover[a]	9.7	
	Smooth brome[b]	15.2	
	Alfalfa[b]	11	
	Western wheatgrass[c]	9.7	
	Alsike clover[c]	11	
	Barley[d]	16.5	
	Japanese millet[d] (added to above mixtures)[d]	8.6	

[a-d] Species with the same superscript letter represent a seeding mixture.

to excellent, with moisture stress and nutrient availability being the controlling factors (Peterson, et al., 1982). Yield fluctuations in corn were not related to heavy metal concentrations in the plants (Hinesly et al., 1979a). In another study, an increase in corn production of 2666 kg ha⁻¹ was observed when Chicago sludge was applied to stripmine spoils at 56 Mg ha⁻¹ (Blessin and Garcia, 1979). Spoil pH appeared to be the major factor governing yield of corn and soybeans on Kentucky mine sites amended with 15 to 120 Mg ha⁻¹ of sludge. Some varieties of soybeans yielded twice that of others; corn varietal differences were not as dramatic (Feuerbacher et al., 1980).

In 1984, Hinesly and Redborg reported on a three-year study where digested sewage sludge was dredged from a storage basin, dewatered and applied on calcareous stripmine spoil at rates of 0, 224, 448, and 896 Mg ha^{-1}. Nine species of grass were seeded on the plots, followed by a planting of corn. They reported that total concentrations of P, Ca, Mg, Fe, Zn, Cd, Cr, Cu, Ni, and Pb were significantly increased in the spoil in proportion to the sludge loading rates. Concentrations of K and Na were decreased in spoil by sludge additions, while concentrations of Mn remained unchanged. Concentrations of all elements remained rather constant in spoil surface samples collected during a 31-month period. Results showed that N, P, Zn, Cd, Cr, Cu, Ni, and Pb added as constituents of sludge remained in the 0 to 30 cm depth of spoil. Changes in K, Na, Ca, Mg, and Fe concentrations resulting from sludge applications were restricted to the 0 to 15 cm depth of spoil. Only pH was affected at spoil depths greater than 30 cm. The pH of the spoil at 30 to 46 cm was decreased by one pH unit with the highest sludge application rate.

In the first two years, corn grain yields were highest on the plots treated with either 224 or 448 Mg ha^{-1} of sludge. But during the third year, the highest corn grain yields were obtained at the highest sludge application rate (3.90 Mg ha^{-1}). Other yields were 1.07, 3.04, and 4.99 for the 0, 224, and 448 Mg ha^{-1} sludge applications, respectively. Stover yields were also the highest with the two higher sludge application rates. Stover yields were 3.49, 4.24, 5.78, and 5.53 Mg ha^{-1} for the 0, 224, 448, and 896 Mg ha^{-1} sludge application rates.

Sludge applications did not affect the concentrations of Mg, Ca, P, Fe, Cr, and Pb in corn tissue samples. However, tissue concentrations of N, Zn, Mn, Ni, Cd, and Cu were increased. These increased element concentrations in the third year were about the same as they were the first year. All corn leaf concentrations of Zn, Mn, Ni, Cr, Pb, and Cu were all below the suggested tolerance level for agronomic crops as reported by Melsted (1973). Only Cd corn leaf concentrations at 7.77 mg kg^{-1} exceeded the suggested tolerance level of 3 mg kg^{-1}.

In a greenhouse study, Tunison and others (1982) grew highbush blueberries using a low-metal sludge from Waynesburg, PA. Without composted sludge, the plants were severely chlorotic and mortality was high, but using composted sludge resulted in healthier plants and increased berry production.

In Kentucky, Powell et al. (1986) investigated the use of organic amendments for the restoration of reconstructed prime farmland following strip mining. Dried sewage sludge and Real Earth™ (a commercially available, dried sewage sludge/garbage compost) were applied at rates of 22.4 and 44.8 dry Mg ha^{-1} to plots, along with 3.4 Mg ha^{-1} of lime, on reconstructed prime farmland and incorporated. Grain sorgham and corn were grown on the plots. Yield of grain sorgham was highest for the sludge-amended plots, followed by control, followed by Real Earth™ applications. Yield of corn was highest for the sludge-amended plots, followed by Real Earth™ and then the control. The organic matter amended soils were less acidic than the nonamended soils and had increased organic matter content, higher soil N, and higher extractable P levels. Water-holding capacity and bulk density of the soil was unaffected by the organic matter amendments.

Trees

While ground cover is the crucial element in initial site stabilization, the potential of woody species for use in sludge management schemes is great, due to their relatively small input into the human food chain and their ability to differentially accumulate metals in specific plant organs (Schneider et al., 1981). In general, establishment of woody vegetation is enhanced by sludge. At the Palzo tract in Illinois (Roth et al., 1982), where 12 tree species were planted, the survival rate was 53% with sludge compared to 19% on the untreated area. On Illinois stripmines and Pennsylvania burned anthracite refuse, hardwood species survival was greater when trees were planted along with herbaceous vegetation, but conifers did better without herbaceous cover (Kardos et al., 1979). Ground cover competition with conifers usually occurs when cover species are planted along with trees, but Berry (1982) found no weed competition problems with conifer establishment when sludge was applied at 34 Mg ha^{-1} on barren land in the Southeast. Grass planting should be delayed until seedlings are established. Virginia pine was successfully established at a seeding rate of 1.4 kg ha^{-1} where there was no herbaceous cover (Scanlon et al., 1973), and the height of the trees was increased with increasing applications of composted sludge. The pines grown on spoils amended with ammonium nitrate fertilizer were markedly chlorotic, whereas those grown in composted sludge-amended spoils were green (Scanlon et al., 1973). Hinkle (1982) noted a significant establishment of volunteer hybrid poplar on sludged pyrite mine refuse in Virginia, after an unsuccessful planting of loblolly pine during a period of drought.

Seedlings of four hardwood species and one pine, usually good performers on reclaimed sites, were grown in spoils amended with a high- and a low-metal sludge (Cd 710 vs 7 mg kg^{-1}; Zn 8010 vs 1075 mg kg^{-1}). Growth, vigor, and survival the first year were highest with the high-metal sludge, probably due to the higher amounts of organic matter and soil conditioning, since that sludge was put on at a higher rate. Species performance was in the following order: European alder > red oak > cottonwood > loblolly pine (Schneider et al., 1981).

In Pennsylvania, dewatered and heat-dried municipal sludge was applied at rates of 0, 40, 75, and 150 Mg ha^{-1} to plots on a burned anthracite coal refuse bank (red gob) (Sopper, 1990a). After incorporation, ten species of tree seedlings were planted. The chemical composition of the sludge and amounts of nutrients and trace metals applied were previously given in Tables 8 and 9. Results of tree survival and total height growth measurements for the three growing seasons are given in Tables 22 and 23. Survival was inversely related to the amount of sludge applied. Survival was the highest on the areas receiving no sludge and decreased with increasing sludge application rates. Much of the tree seedling mortality on the sludge-treated plots was probably due to the greater competition from the dense growth of herbaceous vegetation. After three growing seasons, average tree seedling survivals were 66, 46, 30, and 10% for the 0, 40, 75, and 150 Mg ha^{-1} sludge application rates, respectively.

Table 22.   Average Tree Survival and Height
             Growth for All Species Combined

Sludge Application Mg ha⁻¹	First year	Second year	Third year
		Survival (%)	
0	88	74	66
40	59	51	46
75	43	33	30
150	20	13	10
		Height (cm)	
0	28	60	96
40	39	77	111
75	47	96	149
150	46	114	194

Source: Sopper (1990a).

Unlike survival, tree height growth was directly related to the amount of sludge applied. In all three growing seasons, increases in overall tree height (averaged over all species) were observed with increases in sludge application. After three growing seasons, average height growths for all species were 96, 111, 149, and 194 cm for the four sludge application rates, respectively.

The hardwood species were far superior to the conifer species in terms of both survival and height growth. Black locust had the best overall response of all species tested, with 88 percent survival and 209 cm of height growth after the third growing season. Red pine had the poorest response of all species tested, with 17% survival and 28 cm of height growth after the third growing season. Hybrid poplar and European alder had the next best growth responses.

Table 23.   Individual Tree Species Survival and Height Growth

Tree species	Survival (%)			Height (cm)		
	1st Yr	2nd Yr	3rd Yr	1st Yr	2nd Yr	3rd Yr
White pine	66	52	40	11	20	28
Austrian pine	57	37	26	14	24	32
Virginia pine	56	36	27	17	33	59
Red pine	52	29	17	11	18	28
White spruce	66	50	48	23	26	33
Japanese larch	40	28	24	27	44	74
European alder	73	67	62	46	111	170
Hybrid poplar	70	65	60	65	126	214
Black locust	94	91	88	75	151	209
Black walnut	88	60	48	—[a]	—	—

Source: Sopper (1990a).

[a]   Height not measured due to repeated die back of stem.

Table 24.   Average Total Height Growth of Surviving Trees After Three Growing Seasons

| Tree Species | Sludge Application Rate (Mg ha⁻¹) | | | |
	0	40	75	150
			cm	
White pine	38	24	25	20
Austrian pine	35	28	38	—
Virginia pine	64	44	45	—
Red pine	40	19	26	—
White spruce	34	30	26	26
Japanese larch	85	33	—[a]	—
European alder	189	154	141	—
Hybrid poplar	112	256	292	271
Black locust	185	179	190	178

Source: Sopper (1990a).

[a] No surviving trees.

    Individual tree species height growth responses for the four sludge application rates are given in Table 24. Height growth of all of the conifer species were less on the sludge-treated plots. As mentioned before, this was probably due to the increased competition from the herbaceous vegetation. The more rapidly growing hardwood species were less affected. In fact, hybrid poplar growth was greatly stimulated by the sludge applications. Average height and diameter growth of the hybrid poplar after three growing seasons are given in Table 25. There was a general trend of increased growth up to the 75 Mg ha⁻¹ sludge application; at that point, the growth response leveled off with little or no increase at the 150 Mg ha⁻¹ sludge application. This seems to indicate a saturation level after which the additional nutrients supplied by the sludge do not stimulate additional tree growth. In general, tree seedling survival was higher at the lower sludge application rates. However, tree height growth was the greatest at the higher sludge application rates.

    In Colorado, one-year old transplants of fourwing saltbush and mountain big sagebrush were grown for 9 months in pots containing low pH (3.1) copper mine spoil amended with 0, 30, and 60 Mg ha⁻¹ of air-dried municipal sewage sludge (Sabey et al., 1990). Growth of fourwing saltbush was enhanced from 38- to over

Table 25.   Average Height and Diameter Growth of Hybrid Poplar After Three Growing Seasons

Sludge Application (Mg ha⁻¹)	Height (cm)	Diameter (cm)
0	112	1.0
40	256	2.7
75	292	3.3
150	271	3.4

Source: Sopper (1990a).

Table 26.   Tree Seedling Survival Percentage

Species	1S:1FA	2S:1FA	3S:1FA	All Treatments
White pine	86	79	86	83
Austrian pine	79	93	57	76
Virginia pine	50	50	33	44
Red pine	36	79	57	57
Larch	86	79	79	81
Black locust	93	100	65	86
Arnot bristly locust	64	79	65	69
Alder	72	100	100	91
Black cherry	93	72	86	84
Sugar maple	79	72	79	77
Hybrid poplar	100	75	55	77
All species	76	80	69	

Source: Sopper (1989a).

300-fold by the additions of sewage sludge. Growth of big sagebrush was increased over six-fold. Investigators attributed growth response to increased availability of N, P, and K and improved biological and physical properties of the spoil-sludge mixture. The addition of western wheatgrass to some of the pots containing fourwing saltbush caused a decrease in shrub growth, undoubtedly due to competition for nutrients.

In Pennsylvania, Sopper (1989c) reported on the survival and growth of 11 tree species planted on a contaminated zinc smelter site amended with mixtures of sewage sludge and fly ash. Tree species evaluated and tree survival after one growing season are given in Table 26. The highest rate of survival (80%) for all species was found on the 2S:1FA amendment (Table 26). Species survival ranged from a low of 44% for Virginia pine to 91% for European alder. Species with the highest survival rates across all amendments were alder (91%), black locust (86%), black cherry (84%), white pine (83%), larch (81%), and hybrid poplar (77%).

Table 27.   Tree Seedling Height Growth

Species	1S:1FA (cm)	2S:1FA (cm)	3S:1FA (cm)	All Treatments (cm)
White pine	19	15	19	18
Austrian pine	24	17	20	20
Virginia pine	14	18	21	17
Red pine	17	19	17	17
Larch	34	28	58	40
Black locust	55	49	39	47
Arnot bristly locust	46	35	46	42
Alder	59	53	46	52
Black cherry	52	57	61	57
Sugar maple	34	39	38	37
Hybrid poplar	34	27	52	38
All species	35	32	38	

Source: Sopper (1989a)

Figure 5.   General view of barren plot before the sludge-fly ash amendment was applied.

There was little difference in average height growth within a species among the three amendments (Table 27). Larch had the best height growth (40 cm) of the coniferous species and black cherry had the best height growth (57 cm) of the hardwood species. The two species which exhibited the most vigorous growth were larch and alder. A general view of a portion of the devastated mountain slope is shown in Figure 5 and a general view of the herbaceous vegetation and some alder seedlings at the same location, 15 months after the amendments were applied (Figure 6).

Foliar samples were collected on 10 September 1987 from all tree species and analyzed for nutrients (N-P-K), cations (Ca and Mg), and trace metals (Mn, Fe, Al, B, Cu, Zn, Pb, Cd, Ni, Co, and Cr). Nitrogen concentrations in the coniferous species ranged from 1.00 to 1.70% and in the hardwood species from 1.34 to 3.34% across all amendments. In general, the hardwood species had higher concentrations of nutrients and cations than the coniferous species. Within a species, there was little difference in nutrient and cation concentrations among the three amendment mixtures.

In Pennsylvania, Sopper (1991b) reported on the survival and growth of tree seedlings planted on a steep coal refuse bank amended with a mixture of sludge and fly ash. The sludge application was equivalent to 65 Mg ha^{-1}. Survival and height growth of the six tree species planted are given in Tables 28 and 29. Tree seedling survival in 1989 was similar on both the control and sludge-fly ash amended plots except for tree of heaven which had a significantly higher survival rate on the control plots. In 1990 all tree species, except tree of heaven, had a higher survival rate on the control plots. However, only larch survival was statistically significant. The lower survival rate on the amended plots was probably due to the greater competition from the herbaceous vegetation which was

Figure 6.   General view of herbaceous vegetation and alder tree seedlings on a field plot at the end of the 1987 growing season, 15 months after amendment application.

Table 28. Average Tree Seedling Survival

Species	Control 1989 %	Sludge-Fly Ash 1989 %	Control 1990 %	Sludge-Fly Ash 1990 %
Black locust	100	100	100	97
Red oak	100	96	90	60
European alder	90	87	77	67
Catalpa	100	90	100	83
Tree of heaven	100[a]	87	53	87
Japanese larch	43	27	40[b]	27

Source: Sopper (1991b).

[a,b]   Significant effect at $P < 0.05$ and 0.001, respectively. Comparisons were made between treatments for each species for each year.

Table 29.   Average Tree Seedling Height Growth

Species	Control 1989 (cm)	Sludge-Fly Ash 1989 (cm)	Control 1990 (cm)	Sludge-Fly Ash 1990 (cm)
Black locust	108[b]	80	169[b]	112
Red oak	23	23	25	25
European alder	62	53	95[a]	65
Catalpa	48	45	69	61
Tree of heaven	38[b]	29	37	37
Japanese larch	38	37	34	31

Source: Sopper (1991b).

[a,b] Significant effect at *P* <0.05 and 0.01, respectively. Comparisons were made between treatments for each species for each year.

greatly stimulated by the sludge-fly ash amendment. There was a large mortality of tree of heaven on the control plots in 1990 and this may have been due to Mn toxicity as discussed later. Average survival of all species combined was 77% on the control plots and 70% on the amended plots. In 1989, height growth of black locust and tree of heaven were significantly greater on the control plots. However, by 1990 only black locust and European alder had significantly greater height growth on the control plots. The poorer height growth response on the amended plots may again be due to the greater herbaceous vegetation competition. Black locust exhibited the best overall growth response and Japanese larch had the poorest survival and growth response.

Summary

Sludge applications on mine land generally have had a very beneficial effect on the establishment and growth of grass and legume species. It facilitates rapid establishment and vigorous growth. Sludge amended sites generally have a greater percent areal cover of vegetation and greater yields. Individual plants also tend to have better developed root systems. On sites where the vegetation is not harvested, there is a greater recycling of nutrients and a greater accumulation of organic matter. Sludge applications also generally have increased the yields of field crops such as sweet corn, field corn, corn stover, bush beans, soy beans, small grains, and sorghum.

Sludge applications also have aided in the establishment of tree seedlings. Annual height and diameter growth are generally increased. Hardwood species tend to have a better survival rate than conifers because of the herbaceous vegetation competition. Hardwood species such as black locust, European alder, and hybrid poplar perform the best if grass and legume species are seeded simultaneously.

Vegetation Quality

Macronutrients

Although increases in the major plant nutrients, N, P, K, Ca, and Mg, are usually observed for crops grown in sludge-amended spoils (Hinesly et al., 1982; Griebel et al., 1979; Seaker and Sopper, 1982), the effects are not always consistent. For example, Mathias and others (1979) found that foliar P always increased, K decreased, and N was not much different in plants grown on sludged spoils compared to nonsludged controls. Sludges are often deficient in K. Even 224 Mg ha^{-1} of sludge did not increase foliar K concentrations, but there was sufficient native K for plant growth in acidic Maryland spoils (Griebel et al., 1979). In Pennsylvania, where native feldspars and micas provide K, mine spoils have been successfully revegetated with municipal sludge without additional K, and no deficiency symptoms or yield reductions of forages were observed in five growing seasons (Seaker and Sopper, 1983, 1984).

Nitrogen levels in sweetcorn grain grown on sludge-amended sand and gravel spoils were significantly greater than fertilizer controls (Hornick, 1982), but sludge did not affect P, K, Ca, or Mg concentrations. Whole corn and soybean plants grown with sludge on acid spoils had higher levels of N, P, K, and Ca than control plants (Feuerbacher et al., 1980). If treatment differences are small (2.5 cm vs 5.0 cm of sludge), there may be no difference in forage growth response or in concentrations of N and P in the leaves (Franks et al., 1982). Schneider and others (1981) found lower foliar N concentrations in trees grown with sludge compared to those grown with inorganic fertilizers; they attributed the results to increased biomass and consequent N dilution in the leaves. Differences in sludge type and treatment affect the availability of nutrients to plants. For example, where composted sludge is used, total soil N remains high over a long period of time due to its organic nature and slow mineralization, 10 to 20% the first year (Griebel et al., 1979; Sopper and Kerr, 1981), while N applied in liquid sludges is more readily available (up to 40% the first year).

In Illinois, application of sewage sludge (542 Mg ha^{-1}) in combination with lime (89.6 Mg ha^{-1}) was an effective treatment in establishing vegetation growth on acidic coal refuse material (Pietz et al., 1989b). The effectiveness was related to the ability to supply N, P, and some K to the growing plants. Plant analyses for N, P, K, Ca, and Mg indicated that plant nutrients were in the adequate to high range for alfalfa and grasses, except for N and K. Tissue N concentration in a composite sample of grasses and alfalfa the third year after application was 15.0 g kg^{-1}, suggesting N deficiency (Martin and Matocha, 1973). Plant K concentration was 11.3 g kg^{-1}. A K concentration of <20.0 g kg^{-1} for cool season grasses is considered to be deficient (Martin and Matocha, 1973).

Topper and Sabey (1986) observed a linear increase in the tissue N and P of grasses with increasing levels of sludge applications ranging from 14 to 83 Mg ha^{-1} on coal mine spoil with a pH of 6.5. Roberts et al. (1988) also found that tall fescue tissue N increased as the sludge rate increased from 22 to 224 Mg ha^{-1}. Tissue N concentrations increased the second year on all sludge applications and

Table 30.   Description of Strip Mine Sites in Pennsylvania Reclaimed with Sewage Sludge

County	Spoil Type	Sludge Source	Sludge Treatment	Application Rates (Mg ha^{-1})	Years of Data
Venango	Abandoned bituminous	Farrell Franklin Oil City	Activated sludge/ anaerobic digestion liquid or dewatered	0,11,90,184	3
Lackawanna	Anthracite refuse	Scranton	Activated sludge dewatered	80,108	3
Westmoreland	Abandoned bituminous	Philadelphia	Anaerobic digestion dewatered 1:1 composted to sludge cake	0,134	2
Somerset	Topsoiled bituminous	Philadelphia	Anaerobic digestion dewatered composted	0,202	3
Clarion	Abandoned bituminous	Philadelphia	Anaerobic digestion dewatered 1:1 composted to sludge cake	128	3

then decreased slightly the third year after treatment. Authors concluded that the large cumulative N uptake by the grass over the first two growing seasons may have depleted the mineralizable N, thus limiting production the third year. The authors reported that P tissue concentrations were the highest on a plot treated with 56 Mg ha^{-1} sludge. As biomass increased with higher sludge applications (112 and 224 Mg ha^{-1}), P tissue concentrations were diluted. As with N, they found a decrease in the P tissue concentration for the third year. The higher tissue P concentration the first two years after application may be related to the breakdown and release of organic and inorganic P in the sludge. The fixation of P into Fe forms in the mine spoil may also have contributed to the lower tissue concentrations. Plant tissue K content increased in all three years as the sludge application rate increased, but were not significantly different for the higher sludge rates (112 and 224 Mg ha^{-1}). Calcium and Mg are usually not limiting on mine spoils amended with sludge because lime is almost always added to raise the spoil pH. In the above study, tissue Ca and Mg concentrations increased as the rate of sludge addition increased.

Rock phosphate additions increased yields of fescue and birdsfoot trefoil on Maryland spoils in conjunction with one-time sludge applications (Griebel et al., 1979), but after repeated annual applications of sludge in Fulton Co., Illinois, a buildup of P in the soil of 15 times the optimum level reduced soybean yields (Hinesly et al., 1979a). Limestone incorporation along with sludge may increase concentrations of foliar Ca (Mathias et al., 1979). In most cases, the macronutrient concentrations of forages and other crops grown on sludge-amended spoils are within the range of concentrations in forages grown with inorganic fertilizers on agricultural soils.

Seaker and Sopper (1982) investigated macronutrient uptake by plants on five mine sites treated with single applications of sludge, ranging from 11 to 202 Mg

Table 31. **Mean Foliar Concentrations (mg kg⁻¹) of Macronutrients in Two Grass Species (Orchardgrass and Tall Fescue) and One Legume Species (Birdsfoot Trefoil) the Third Year After Sludge Application**

Element	Species									
Applied N (kg ha⁻¹)		0	284	726	968	1165	1241	1437	1691	2388
% N	grasses	1.2	1.4	2.1	2.1	1.3	1.8	1.5	1.8	1.3
	legume	2.6	2.8	3.3	3.7	3.4	2.6	2.0	2.6	3.6
Applied P (kg ha⁻¹)		0	66	328	414	451	464	776	839	1302
% P	grasses	0.19	0.46	0.42	0.45	0.50	0.39	0.43	0.46	0.29
	legume	0.18	0.17	0.18	0.23	0.20	0.25	0.30	0.25	0.11
Applied K (kg ha⁻¹)		0	5	8	17	101	139	176	181	246
% K	grasses	1.8	2.4	2.5	2.4	2.1	2.4	2.7	2.9	2.1
	legume	1.9	1.5	1.4	1.6	1.3	1.4	2.0	1.5	0.77
Applied Ca (kg ha⁻¹)		0	445	893	1809	1816	1915	2926	6214	8522
% Ca	grasses	0.42	0.56	0.53	0.51	0.27	0.44	0.51	0.46	0.40
	legume	1.1	1.6	1.5	1.5	0.25	1.2	1.5	1.6	1.7
Applied Mg (kg ha⁻¹)		0	75	187	370	378	485	507	692	960
% Mg	grasses	0.21	0.25	0.28	0.48	0.27	0.20	0.38	0.39	0.24
	legume	0.27	0.22	0.23	0.44	0.20	0.06	0.41	0.33	0.18

ha⁻¹. A description of the five sites are given in Table 30. The amounts of macronutrients applied in the sludge are given in Table 31. Control vegetation adjacent to sludged sites received only conventional fertilizer and lime applications. Because sludge is applied only once, relatively high levels of nutrients are applied in order to meet vegetation needs for the five-year period which a self-maintaining vegetative cover is required to comply with Pennsylvania reclamation regulations. Foliar concentrations of N, P, K, and Mg were higher in sludge-grown plants than in control plants with one exception, sludge-grown birdsfoot trefoil had lower concentrations of K and Mg than control plants. Sludge-applied Ca appeared to increase birdsfoot trefoil foliar Ca concentrations slightly but had little effect on the grass species. Birdsfoot trefoil had higher foliar N concentrations, as expected for a legume, while the grasses had higher foliar P concentrations. Foliar concentrations of macronutrients did not show a linear response to the amount of element applied.

The following is the normal range of macronutrients found in grass foliage samples analyzed at the Penn State Merkle Lab. Most of samples analyzed at the lab are from normal farming operations on land not disturbed by mining:

Element	Concentration Range (%)
N	1.57–2.94
P	0.16–0.30
K	1.07–2.31
Ca	0.29–0.69
Mg	0.10–0.22

Foliar concentrations of all macronutrients were at levels which would indicate sufficient nutrients were available to support optimum growth. Most macronutri-

Table 32.   Average Concentrations of Nutrients in the Foliage
of the Herbaceous Species

Species	Nutrient	Control (%)	2S:1FA (%)
Birdsfoot trefoil	N	3.37	3.62
	P	0.19	0.22
	K	1.14	1.41
Crownvetch	N	2.31	2.47
	P	0.16	0.22[a]
	K	1.29	1.53[a]
Tall fescue	N	1.45	2.76[b]
	P	0.29	0.31
	K	1.91	1.98
Orchardgrass	N	1.66	2.50[c]
	P	0.27	0.37[d]
	K	2.34	2.16

Source: Sopper (1991b).

[a,b,c]   Significant effect P <0.05, 0.01 and 0.001, respectively.

ent values in both grasses and legume on all the mine sites treated with sludge were generally on the high side of the range given above.

In Pennsylvania, Sopper (1991b) reported on the uptake of nutrients by herbaceous vegetation growing on a steep coal refuse bank amended with a mixture of sludge and fly ash. The sludge application was equivalent to 65 dry Mg ha^{-1}. Details of this study have already been discussed.

Average concentrations of nutrients in the foliage of the herbaceous vegetation in 1990 are given in Table 32. Foliar nutrient concentrations, in general, were higher in plants growing on the sludge-fly ash amended plots. Tall fescue and orchardgrass both had significantly higher foliar N concentrations on the amended plots. Crownvetch and orchardgrass also had significantly higher foliar P concentrations on the amended plots. Crownvetch was the only herbaceous species that had a higher foliar K concentration on the amended plots.

Table 33.   Average Concentrations of Nutrients in the Foliage of the Tree Species

Species	Plot	N (%)	P (%)	K (%)
Catalpa	Control	1.44	0.16	1.45[a]
	2S:1FA	2.30	0.15	1.28
Alder	Control	2.20	0.20	1.47[a]
	2S:1FA	2.11	0.16	1.12
Tree of heaven	Control	1.57	0.21	1.72
	2S:1FA	2.16	0.23	1.76
Red oak	Control	1.88	0.27	0.72
	2S:1FA	2.21	0.29	0.83
Black locust	Control	3.58	0.16	1.38
	2S:1FA	3.64	0.20	1.49
Larch	Control	1.30	0.20	0.78
	2S:1FA	1.68[b]	0.25[a]	0.77

Source: Sopper (1991b).

[a,b]   Significant effect at P <0.05 and 0.01, respectively.

Average concentrations of nutrients in the foliage of the tree species in 1990 are given in Table 33. Concentrations of N and P were generally higher in the foliage of the tree species growing on the sludge-fly ash amended plots. Larch was the only tree species which had significantly higher concentrations of N and P on the sludge-fly ash amended plots. Foliar K concentrations were quite similar for most tree species, except that catalpa and alder had significantly higher foliar K concentrations on the control plots.

Sludge applications on mine land generally increase the total nitrogen concentration in the foliage of vegetation. Excessively high concentrations of nitrogen in plants are not detrimental to the plant but may lead to high nitrates in the plant, which could cause metabolic disruptions in a foraging animal. When large amounts of nitrate nitrogen are ingested by a ruminant animal such as a cow, the bacteria within the rumen reduce nitrate to nitrite. The nitrite produced is subsequently absorbed through the intestine walls and into the bloodstream. In the bloodstream, nitrite, which is a strong oxidizing agent, converts ferrous iron in the center of the hemoglobin molecule to ferric iron greatly diminishing the ability of blood to transport oxygen to the sites in the animal's body requiring it for normal metabolic activity. As a result, animals may exhibit symptoms of oxygen starvation with bluish discoloration of mucous membranes and difficulty in breathing. Severe cases of nitrate toxicity may result in a rapid decline in the animal's condition, possible abortion of pregnant animals, a drop in milk production, a decline in rate of weight gain, and eventually death due to a deficiency of oxygen in the blood (Follett et al., 1981). This author was unable to find any published documentation of this phenomena occurring as a result of sludge applications on mine land.

Trace Metals

It is generally agreed that municipal sludge improves the capacity of spoil material to support vegetation, but questions often arise about the uptake of trace metals from sludges by forages, crops, trees, and other plants, which may be involved in food-chain dynamics.

Grass and Legume Species. Several studies indicate that a decrease in trace metal concentrations of sludge-grown vegetation over time may occur where a single application of sludge is used. Metal concentrations in tall fescue from Ohio mine spoils amended with up to 716 Mg ha^{-1} of sludge were considerably lower in the third growing season than they were in the first (Haghiri and Sutton, 1982). On anthracite refuse in Pennsylvania, Cu, Zn, and Cd increased in reed canarygrass tissues the first growing season after sludge was applied, but by the second and third years, with few exceptions, metal concentrations decreased to control levels or below (Kerr et al., 1979). As part of an extensive demonstration program in Pennsylvania, tall fescue, orchardgrass, birdsfoot trefoil, and crownvetch from two sludge-amended coal sites were analyzed for seven trace metals over a five-year period (Seaker and Sopper, 1983, 1984). Results showed a definite decrease in heavy metal concentrations over time. With few exceptions, Cu, Zn, Cr, Pb, Co,

Table 34.   Suggested Permissible
Tolerance Levels of Trace
Metals in Agronomic Crops

Element	Suggested Tolerance Level (mg kg^{-1})
Fe	750
Mn	300
Al	200
Zn	300
Cu	150
B	100
Cr	2
Pb	10
Co	5
Ni	50
Cd	3

Source: Council for Agricultural Science and
Technology (1976); Melsted, (1973), and
University of Georgia Cooperative Ex-
tension Service (1979).

Cd, and Ni remained well below suggested tolerance levels for agronomic crops as shown in Table 34.

Seaker and Sopper (1982) found the same trend occurring on five mine sites treated with single applications of sludge, ranging from 11 to 202 Mg ha^{-1}. A description of the five sites are given in Table 30. The amounts of metals applied in the sludge and foliar concentrations after three growing seasons are given in Table 35. Control vegetation adjacent to sludged sites received only conventional fertilizer and lime applications. Foliar concentrations of Fe, Al, Zn, Co, Pb, Ni, and Cd in sludge-grown plants were all below the suggested tolerance level cited by Melsted (1973) and were consistently within 5 mg kg^{-1} of control plant concentrations. Foliar Cu concentrations were consistently within the average range for agronomic crops (Melsted, 1973; Chapman, 1982; University of Georgia, 1979). Foliar Zn concentrations were higher in sludge-grown plants than in control plants; however, the concentrations were still within the average range for agronomic crops. Foliar Fe and Mn concentrations in sludge-grown plants were generally lower than in control plants and were mostly within the average range for agronomic crops. Cobalt is essential for N-fixing bacteria, and thus for legume species. Foliar Co concentrations in sludge-grown plants were seldom higher than 2 mg kg^{-1} of control plant foliar concentrations. It appears that the sludge applications supplied adequate, but not excessive, amounts of this essential element to the plants. Sludge applied Pb had little effect on foliar Pb concentrations. Foliar Pb in sludge-grown plants fluctuated above and below foliar Pb concentrations in control plants, but never exceeded control values by 20 mg kg^{-1} and were consistently within the average range for agronomic plants. Lead poses little hazard to crops because it is usually immobilized by the high amounts of phosphate in sludge (Chaney, 1973), and soil Pb would probably have to reach 10,000 mg kg^{-1} for significant effects on plant to occur (Council for Agricultural Science and Technology, 1976). Foliar Ni and Cr concentrations in sludge-grown plants were

Table 35.   Mean Foliar Concentrations (mg kg⁻¹) of Metals in Two Grass Species (Orchardgrass and Tall Fescue) and One Legume (Birdsfoot Trefoil) the Third Year After Sludge Application

Element	Species								
Applied Mn (kg ha⁻¹)	0	9	110	112	121	131	148	153	300
Mn (mg kg⁻¹)	grasses 279	90	274	41	188	146	298	72	351
	legume 222	66	28	43	244	121	132	50	127
Applied Fe (kg ha⁻¹)	0	261	319	2501	2662	3389	3430	3794	5390
Fe (mg kg⁻¹)	grasses 445	79	78	93	64	97	74	53	53
	legume 199	95	105	67	110	95	72	25	98
Applied Al (kg ha⁻¹)	0	219	514	550	704	1113	1325	1673	2103
Al (mg kg⁻¹)	grasses 304	35	48	31	36	29	48	65	30
	legume 137	34	39	35	39	29	41	48	11
Applied Zn (kg ha⁻¹)	0	21	64	72	86	147	183	245	342
Zn (mg kg⁻¹)	grasses 21	25	27	35	26	65	62	37	50
	legume 29	33	32	46	35	42	114	43	63
Applied Cu (kg ha⁻¹)	0	21	63	67	76	92	129	131	148
Cu (mg kg⁻¹)	grasses 9.1	7.3	7.5	10.0	11.8	7.8	8.8	11.4	10.9
	legume 8.4	8.3	8.9	6.3	10.6	6.8	9.2	8.9	9.8
Applied Cr (kg ha⁻¹)	0	16	16	21	36	40	42	42	74
Cr (mg kg⁻¹)	grasses 2.3	7.7	0.25	0.25	11.6	0.25	<0.01	<0.01	2.5
	legume 1.5	1.0	0.33	0.33	8.3	0.75	<0.01	<0.01	1.7
Applied Co (kg ha⁻¹)	0	0.2	0.4	0.8	0.9	1.1	2.0	2.2	3.0
Co (mg kg⁻¹)	grasses 1.5	0.42	0.04	1.7	0.13	2.2	1.2	1.1	0.42
	legume 1.7	0.75	1.3	3.3	<0.01	2.4	0.42	1.7	0.25
Applied Pb (kg ha⁻¹)	0	10	27	49	55	59	67	80	131
Pb (mg kg⁻¹)	grasses 2.6	3.2	2.5	1.4	2.7	1.5	0.71	1.5	1.5
	legume 3.7	3.3	1.8	3.9	1.8	3.3	2.4	2.0	2.2
Applied Ni (kg ha⁻¹)	0	1	4	6	7	8	12	13	30
Ni (mg kg⁻¹)	grasses 5.3	10.9	2.5	2.7	16.0	1.6	5.3	1.4	4.5
	legume 7.8	4.5	4.4	4.3	15.2	1.3	6.3	4.4	6.0
Applied Cd (kg ha⁻¹)	0	0.1	0.2	0.2	0.6	0.6	1.2	1.7	5.0
Cd (mg kg⁻¹)	grasses 0.27	0.06	0.12	0.44	0.14	1.1	0.16	0.15	1.3
	legume 0.23	0.03	0.11	0.57	0.04	1.3	0.04	0.16	1.4

Source: Seaker and Sopper (1982).

generally the same or lower than in control plants. On sites were Ni and Cr foliar concentrations were higher than controls in the third year, they decreased below control plant values in the fourth year. Foliar Cr concentrations only exceeded the suggested tolerance level (2 mg kg⁻¹) in the control grass species (2.3 mg kg⁻¹) and in the sludge-grown plants (2.5 mg kg⁻¹) on the highest sludge application (74 kg ha⁻¹ Cr applied). Foliar Cd concentrations were generally lower in sludge-grown plants than in control plants and were within the average range for agronomic crops on five out of the eight sites. The highest Cd application on the strip mine sites was 5 kg ha⁻¹. Several studies have indicated that at rates of 2 to 29 kg ha⁻¹ yr⁻¹ Cd, most of the metal was converted to forms unavailable to the plant (Council for Agricultural Science and Technology, 1976).

Research has shown that trace element activity, and thus plant uptake of trace metals from sludged soils, tend to approach a maximum as the sludge rate increases (Corey et al., 1987). The data in Table 35 generally show the plateau effect. For instance, foliar concentrations of Pb in both grasses and the legume were not much different where sludge applications applied 10 kg ha^{-1} and 131 kg ha^{-1}. Trace metal uptake by plants is not a linear response to the amount of sludge, and thus the amount of metal, applied. Sludge contains iron and humic material that strongly adsorbs metals. Any increase in uptake at the plateau level is dependent on the sludge-constituent concentration and the ability of the sludge to adsorb the constituent.

Other studies show no decline in metal concentrations, but these are often cases where sludge was applied annually for a number of years. In Illinois, metals in forage did not appreciably change, even after four successive years of Chicago sludge applications, although values were higher than on control plots (Fitzgerald, 1982). Hinesly et al. (1982) believe Cu and Zn may become less available with time because plant uptake decreased after sludge applications were terminated.

Several factors may account for decreasing metal concentrations over time. Iron and phosphorus added in sludge may complex with metals, forming sparingly soluble precipitates (Cunningham et al., 1975). Metals may bind with the humic fraction of sludge in the spoil-sludge mixture (Haghiri and Sutton, 1982). Although composted sludge additions increased Cu and Ni in forages after two years growth on an abandoned stripmine, the foliar concentrations decreased as compost application increased from 56 to 224 Mg ha^{-1} due to increased pH. When rock phosphate or limestone and sludge were added, uptake of Ni, Cu, Zn, and Cd decreased (Griebel et al., 1979). In another study, increasing the sludge rate from 314 to 627 Mg ha^{-1} resulted in a decrease of Mn, Zn, Ni, and Cd in tall fescue and alfalfa over two years. Cd and Zn reductions may be attributed to increased pH, and Mn and Ni reductions to the additional organic matter supplied. Another important factor involved in metal concentrations in plant tissues is the possible "dilution" effect that occurs as a result of increased biomass production where sludges are used (Kerr et al., 1979). The sampling date determines in part the concentrations of metals in plants, and seasonal differences may be greater than treatment differences. For example, in one study Pb concentrations of tall fescue were four times higher in January than they were the previous October (Mathias et al., 1979). Forages collected in the fall had Mn and Al concentrations high enough to be considered potentially toxic, but the following June the levels were safe for animal rations (Sutton and Vimmerstedt, 1974). This observation emphasizes the importance of sampling plant material at a predetermined growth stage or time of year. The part of the plant analyzed also influences metal concentration values. For example, significant differences in metal concentrations existed between grain and stover of winter rye for Cu, Pb, Zn, Mn, and Cd, but not for Ni and Cr (Stucky and Bauer, 1979). There are often differences in metal concentrations of different species grown under the same sludge regime, but this is not always the case. Stucky and Bauer (1979) observed little difference in Cu, Zn, Ni, Cd, Cr, Mn, or Pb among switchgrass, orchardgrass, and tall fescue, and Seaker and Sopper (1982) found no consistent differences in concentrations of the same elements in tall fescue, orchardgrass, and birdsfoot trefoil. McBride and others

(1977), however, reported definite differences in Cd uptake between rye, Sudan grass, tall fescue, and reed canarygrass.

In general, sludge-grown vegetation usually contains higher concentrations of metals like Cu, Zn, Cr, Pb, Cd, and Ni than do plants grown without sludge, but the heavy metal increases are of varying degrees, and in most cases are not great enough to be a threat to human or animal health.

Within three growing seasons on sludge-amended spoils in Pennsylvania (Seaker and Sopper, 1982), concentrations of six metals in forages were continually below the suggested tolerance levels for agronomic crops (Council for Agricultural Science and Technology, 1976, 1980; Melsted, 1973). Annual mean concentrations from five sites over two to five years showed Cu and Zn concentrations in the forage below the required concentrations for total dairy rations. In other studies, Mn and Al accumulated to potentially harmful levels in forages the first year after sludge, but decreased to safe levels at the start of the second growing season (Franks et al., 1982). Cadmium concentrations in tall fescue grown with 224 Mg ha^{-1} sludge reached 2.3 mg kg^{-1} vs 0.5 mg kg^{-1} suggested as a safe food level (John et al., 1972; Mathias et al., 1979). On two sites near zinc smelter operations in Oklahoma, Cu concentrations of sludge-grown grasses were consistently below suggested tolerance levels, but Cd concentrations were usually in the yield-reducing range. The major source of metals, however, may not have been the sludge but rather the contaminated soil (Franks et al., 1982). In most cases, plant toxicity symptoms do not appear and metals are below suggested tolerance levels (Stucky and Newman, 1977; Seaker and Sopper, 1982; Hinkle, 1982; Scanlon et al., 1973).

Manganese is an element of concern in mine spoils since it may be accumulated by plants to phytotoxic levels. Manganese can become toxic to tall fescue growth in mine spoils when pH decreases to <5 (Palazzo and Duell, 1974). Roberts et al. (1988) found that tall fescue tissue Mn levels were highest the first year after applying sludge at rates from 22 to 224 Mg ha^{-1}, but were not at phytotoxic levels. Tissue Mn levels then decreased after the first year and as sludge rates increased. The authors concluded that the sludge additions increased fescue production and appear to have diluted tissue Mn levels. They also reported that fescue tissue Zn, Cu, and Cd concentrations increased as the sludge rate increased. Concentrations of Zn in fescue tissue decreased in treatments amended with ≥56 Mg ha^{-1} sludge over the two years following application. Cadmium tissue concentrations were the highest the first year, but well below phytotoxic levels, and then decreased with time.

In Pennsylvania, Sopper (1990a) reported on the uptake of trace metals in reed canarygrass and crownvetch seeded on an anthracite burned refuse bank amended with 0, 40, 75, and 150 Mg ha^{-1} of dewatered and heat-dried sludge. Sludge applications had little effect on the trace metal concentrations in the foliage of reed canarygrass.

Significant increases were only found for Zn in the first year on the 150 Mg ha^{-1} and in the third year on the 75 and 150 Mg ha^{-1} plots (Table 36). Significantly higher amounts of Cd were also found in the foliage of plants on the 150 Mg ha^{-1} the first year. Foliar concentrations of Cu were similar on all

Table 36. Average Concentrations of Trace Metals in Foliar Samples of Reed Canarygrass

Sludge Application (Mg ha⁻¹)	Tolerance Level (mg kg⁻¹)	First year (mg kg⁻¹)	Second year (mg kg⁻¹)	Third year (mg kg⁻¹)
		Cu		
0	150	6.9 a[a]	9.9 a	9.6 a
40		13.9 a	10.1 a	9.0 a
75		16.9 a	12.8 a	11.2 a
150		18.9 a	15.4 a	12.3 a
		Zn		
0	300	33.2 a	48.0 a	37.4 a
40		40.0 a	46.0 a	47.0 ab
75		42.8 a	61.6 a	59.8 bc
150		79.2 b	69.3 a	65.6 c
		Cd		
0	3	0.23 ab	0.23 a	0.16 a
40		0.14 b	0.14 a	0.06 a
75		0.36 a	0.18 a	0.12 a
150		0.54 c	0.23 a	0.11 a
		Ni		
0	50	9.2 a	6.9 a	7.4 a
40		5.5 a	3.3 b	3.8 a
75		6.1 a	4.0 b	4.1 a
150		6.2 a	4.1 b	3.6 a
		Pb		
0	10	10.2 a	15.0 a	12.1 a
40		5.1 b	10.5 a	15.5 a
75		4.5 b	7.3 a	8.5 a
150		4.7 b	10.5 a	9.3 a

Source: Sopper (1990a).

[a] Means of the four sludge treatments followed by the same letter are not significantly different at the 0.05 level of significance.

plots and concentrations of Ni and Pb were generally higher on the control plots (0 Mg ha⁻¹).

The foliar trace metal concentrations resulting from the sludge applications were minimal and well below the suggested tolerance level reported by Melsted (1973). The tolerance level is not a phytotoxic level but suggests a foliar concentration level at which one may expect to get decreases in crop yields.

Similarly, there were no significant increases in foliar trace metal concentrations of Penngift crownvetch except for Ni the second year on the 75 and 150 Mg ha⁻¹ plots (Table 37). Again, the increases were minimal and well below the suggested tolerance level. No phytotoxic symptoms were observed on any vegetation.

Table 37.   Average Concentration of Trace Metals in Foliar Samples of Penngift Crownvetch

Sludge Application (Mg ha⁻¹)	Tolerance Level (mg kg⁻¹)	First Year (mg kg⁻¹)	Second Year (mg kg⁻¹)	Third Year (mg kg⁻¹)
		Cu		
0	150	8.2 a[a]	12.4 a	11.2 a
40		9.4 a	11.0 a	9.4 a
75		11.1 a	13.3 a	11.0 a
150		12.0 a	23.1 a	13.3 a
		Zn		
0	300	71.5 a	77.6 a	89.2 a
40		41.0 a	43.3 b	52.9 b
75		49.1 a	49.5 b	38.2 b
150		63.2 a	77.0 a	61.8 ab
		Cd		
0	3	0.37 a	0.49 a	0.44 a
40		0.03 a	0.22 a	0.32 a
75		0.06 a	0.30 a	0.28 a
150		0.13 a	0.49 a	0.31 a
		Ni		
0	50	11.3 a	2.4 a	19.1 a
40		5.1 a	3.4 ab	5.5 b
75		3.6 a	5.9 c	6.1 b
150		6.8 a	4.6 bc	5.6 b
		Pb		
0	10	10.2 a	8.9 a	7.9 a
40		8.9 a	9.3 a	10.1 a
75		12.9 a	11.2 a	8.6 a
150		9.2 a	15.0 a	9.5 a

Source: Sopper (1990a).

[a]   Means of the four sludge treatments followed by the same letter are not significantly different at the 0.05 level of significance.

Pietz et al. (1989b) reported the successful establishment of a vegetative cover on acidic coal refuse treated with sludge, lime, and gypsum and various combinations of each. Sludge application rate was 542 Mg ha⁻¹. Concentrations of Cd, Zn, Cu, and Ni in composite plant samples ranged from 1.3 to 22.8, 62 to 758, 1.8 to 18.8 and 4.4 to 33.8 mg kg⁻¹, respectively.

The highest values for Cd and Zn exceeded the tolerance levels (3 mg Cd kg⁻¹ and 300 mg Zn kg⁻¹) reported in the literature (Melsted, 1973; CAST, 1976). The high values were generally associated with the treatments that included gypsum. Authors also reported that concentrations of Al, Fe, and Pb in plant tissue from some treatments exceeded phytotoxic limits. Aluminum and Fe concentrations in most treatments exceeded the limits of 200 and 750 mg kg⁻¹ proposed by Melsted (1973), respectively.

Joost et al. (1987) also treated coal refuse material with various sludge application rates (225 to 900 Mg ha⁻¹) and examined trace metal uptake by forage grass herbage. They reported that although Cd, Cr, Pb, and Ni were present in the coal refuse at levels considered toxic to plants, there appeared to be no detrimental effects on the grasses grown. In fact, tissue accumulation of all trace metals was within the range considered safe for animal consumption (Underwood, 1971). Highest concentrations of Cd, Cr, Pb, and Ni in forage grass harvested over a four-year period on a plot treated with 900 Mg ha⁻¹ of sludge were 9, 11, 26, and 10 mg kg⁻¹, respectively. The maximum suggested safe in feed is 160, 50, 300, and 1100 mg kg⁻¹, respectively (Underwood, 1971; Church and Pond, 1974).

Kiefer et al. (1983) investigated the effects of applications of sewage sludge, fly ash, and chicken manure on the chemical composition of plants and soil on three abandoned minesites in West Virginia. Sewage sludge applications were 0, 45, and 90 Mg ha⁻¹. Alfalfa and sericea lespedeza were seeded on the plots. Sewage sludge and chicken manure did not affect plant growth. Sludge applications increased the P, Ca, Zn, Cu, Cr, Ni, Pb, and Cd concentrations in the mine soils.

Sopper (1989a) reported on the effects of sludge-fly ash amendments applied to a zinc smelter contaminated site on vegetation quality. Background details on this project have already been described. Foliar samples of all seeded species were collected for analyses of nutrients and trace metals. Nutrient concentrations were quite adequate for optimum growth. For the grass species, foliar N concentrations ranged from 1.55 to 3.97%, foliar P ranged from 0.11 to 0.44%, and foliar K ranged from 1.15 to 2.65% across all amendments. Foliar concentrations of P and K in the legume species were similar; however, N concentrations were generally higher than in the grass species. Foliar concentrations of Cu, B, and Ni were all below the suggested tolerance level in all species on all three amendment mixtures except for B in birdsfoot trefoil on the 1S:1FA mixture (Table 38). The tolerance levels shown are not phytotoxic levels but suggest foliar concentration levels at which one may expect decreases in growth (Melsted, 1973; Adriano, 1986). Foliar Zn, Cd, and Pb concentrations exceeded the tolerance level in all species on all three amendment mixtures except for Cd in ryegrass on the 3S:1FA mixture. Highest Zn concentrations were found in ryegrass, big bluestem, and orchardgrass. Big bluestem also had the highest foliar Pb concentrations. Tall fescue had the highest foliar Cd concentrations. Foliar Mn concentrations exceeded the tolerance level in crownvetch and Lathco flatpea on all amendments, whereas concentrations in birdsfoot trefoil were lower than the tolerance level on all amendments. All grass species had foliar Mn concentrations under the tolerance level except for ryegrass on all amendments, tall fescue on the 2S:1FA amendment, and orchardgrass on the 1S:1FA and 3S:1FA amendments. No phytotoxicity symptoms were observed on any species. A comparison of 1986 and 1987 foliar analyses results indicated that concentrations of B, Cu, and Pb generally increased in all grass and legume species and Mn, Zn, Cd, and Ni generally decreased. Concentrations of Ni were variable, generally increasing in the legumes and decreasing in the grasses.

Table 38.   Foliar Trace Metal Concentrations (mg kg⁻¹) in Herbaceous Species

Species	Amendment Mixture	Mn	B	Cu	Zn	Pb	Cd	Ni
Blackwell	1S:1FA	108	11	16	576	30	4	2
switchgrass	2S:1FA	189	22	18	715	83	5	3
	3S:1FA	(no sample)						
Cave-in-rock	1S:1FA	(no sample)						
switchgrass	2S:1FA	135	18	25	961	113	8	3
	3S:1FA	35	8	14	491	76	6	2
Niagara big	1S:1FA	85	30	27	1066	87	11	5
bluestem	2S:1FA	98	15	22	1089	118	7	2
	3S:1FA	92	13	37	808	51	16	8
Tall fescue	1S:1FA	101	16	17	721	49	20	4
	2S:1FA	379	28	24	950	69	20	7
	3S:1FA	161	20	25	762	43	21	8
Ryegrass	1S:1FA	333	29	50	1115	60	15	14
	2S:1FA	632	36	36	1396	51	17	20
	3S:1FA	499	29	46	804	87	1	16
Wheatgrass	1S:1FA	94	14	24	605	75	14	3
	2S:1FA	168	25	21	470	38	8	3
	3S:1FA	92	13	27	451	60	9	3
Orchardgrass	1S:1FA	331	36	20	550	57	7	7
	2S:1FA	125	18	26	1582	98	17	9
	3S:1FA	360	54	30	474	64	5	9
Lathco flatpea	1S:1FA	375	40	48	979	63	14	16
	2S:1FA	694	63	45	2082	87	17	12
	3S:1FA	438	55	41	1080	56	12	11
Birdsfoot trefoil	1S:1FA	295	128	34	666	56	11	11
	2S:1FA	250	94	36	1121	59	26	12
	3S:1FA	216	79	35	588	56	16	11
Crownvetch[a]	1S:1FA	392	81	24	1356	62	18	13
Tolerance level[b]		300	100	150	300	10	3	50

Source: Sopper (1989a).

[a]   No plants available to sample on the 2S:1FA and 3S:1FA.
[b]   Melsted, (1973).

Table 39.   Average Concentrations of Trace Metals in the Foliage of the Herbaceous Species

Species	Plot	mg kg⁻¹						
		Mn	Cu	B	Zn	Pb	Cd	Ni
Birdsfoot	C	103	9	17	41	6	0.19	11.5[c]
Trefoil	2S:1FA	113	14[b]	67[a]	53[a]	10[a]	0.48[a]	5.3
Crownvetch	C	96[b]	7	11	53	4	0.11	4.9
	2S:1FA	86	13[b]	40	76[a]	8[a]	0.56	6.7[a]
Tall fescue	C	62	5	5	10	2	0.17	5.6
	2S:1FA	59	12[a]	9	28[b]	4[b]	0.47	4.2
Orchardgrass	C	320[b]	6	9	31	3	0.19	6.1
	2S:1FA	73	13	8	32	4	0.18	3.5
Suggested tolerance level[d]		300	150	100	300	10	3	50

Source: Sopper (1991b).

[a,b,c]   Significant effect at $P < 0.05$, 0.01, 0.001, respectively.
[d]   Melsted (1973).

Table 40. Chemical Analysis of Dewatered Sludge Applied and Amounts of Elements Applied at 184 Mg ha⁻¹ Rate (Dry Weight Basis) on the Bituminous Strip Mine Spoil Bank

Constituent	Average Concentration (mg kg⁻¹)	Amount Applied (kg ha⁻¹)
Total P	4624	918
Total N	12188	2388
K	93	18
Ca	9970	1834
Mg	2082	383
Zn	811	147
Cu	661	129
Pb	349	55
Ni	69	12
Cd	3.2	0.6
pH	7.9	

Source: Sopper and Seaker (1990).

More recently, Sopper (1991b) reported on the trace metal uptake of herbaceous vegetation growing on a steep coal refuse bank amended with a mixture of sludge and fly ash. The sludge application was equivalent to 65 dry Mg ha⁻¹.

Average concentrations of trace metals in the foliage of the herbaceous species are given in Table 39. Trace metal concentrations were, in general, higher in plants growing on the amended plots. Trace metal concentrations in all species on both plots were below the suggested tolerance level, except for Mn in orchardgrass. Both crownvetch and orchardgrass had significantly higher foliar Mn concentrations on the control plots. Birdsfoot trefoil, crownvetch, and tall fescue all had significantly higher foliar concentrations of Cu, Zn, and Pb on the amended plots. Birdsfoot trefoil was the only species which had a significantly higher foliar Cd concentration on the amended plots. Foliar Ni concentrations in birdsfoot trefoil were significantly higher in plants growing on the control plots.

Little information is available on long-term effects of single applications of sludge on mine land and metal uptake by vegetation. Issues that often arise are (1) what happens after all the sludge has been mineralized and all the nutrients and trace metals have been released to the soil and are potentially available for leaching and plant uptake, and (2) will the vegetation persist or deteriorate? However, two recent publications by Sopper and Seaker (1990) and Sopper (1991) shed some light on this subject.

In the first study, they resampled orchardgrass and crownvetch growing on an abandoned strip mine spoil bank that had been amended with 184 Mg ha⁻¹ of dewatered municipal sludge 12 years earlier. The sludge and agricultural lime (12.3 Mg ha⁻¹) were applied on the site in 1977 and incorporated. The average concentrations of nutrients and trace metals applied in the sludge are given in Table 40. The amounts of trace metals applied are given in Table 41 along with the U.S. Environmental Protection Agency (EPA) and Pennsylvania Department of Environmental Resources (PDER) interim guideline recommendations (U.S.

Table 41.   Trace Metal Loadings of the Sludge Application and Lifetime Loadings Recommended by the EPA and PDER

| | Trace Metal Loadings (kg ha⁻¹) | | |
Constituent	Sludge application 184 Mg ha⁻¹	EPA[a] (CEC 5-15)	PDER
Cu	129	280	112
Zn	147	560	224
Cr	74	NR[b]	112
Pb	55	800	112
Ni	12	280	22
Cd	0.6	11	3
Hg	0.09	NR[b]	0.6

Source: Sopper and Seaker (1984).

[a]   Average CEC of site ranged from 11.6 to 15.2 meq 100g⁻¹. From U.S. EPA (1983).
[b]   No recommendation given by EPA.

Environmental Protection Agency, 1977; Pennsylvania Department of Environmental Resources, 1977). The amounts of trace metals applied were well below the recommended lifetime limits, except for Cu, which slightly exceeded the PDER guidelines.

Immediately after sludge application and incorporation, the site was broadcast seeded with a mixture of two grasses, Kentucky-31 tall fescue (22 kg ha⁻¹) Pennlate orchardgrass (22 kg ha⁻¹), and two legumes, Penngift crownvetch (11 kg ha⁻¹) and Empire birdsfoot trefoil (11 kg ha⁻¹). Then the site was mulched with straw and hay at the rate of 3.8 Mg ha⁻¹.

The site was completely vegetated by August 1977, three months after sludge application, which has persisted throughout the 12-year period. Average annual dry matter production for the first five years and in 1989 was as follows:

Year	Yield (Mg ha⁻¹)
1977	6.0
1978	9.3
1979	11.3
1980	31.2
1981	22.6
1989	15.5
AHY	4.0

Dry matter production increased during the first four years, leveling off in 1981. In 1989 it was slightly lower but still well above the average hay yield (AHY) for undisturbed farmland soils in the county. During the first two years the two grass species dominated the site, but by the third growing season, the two legume species predominated and persisted through the fifth year (1981). However, by 1989 the birdsfoot trefoil had almost disappeared and now the dominating vegetative cover consists mostly of crownvetch and orchardgrass.

Table 42. **Mean Foliar Concentrations (%) of Macronutrient Elements in Orchardgrass and Birdsfoot Trefoil Collected from the Control Plot and Sludge-Amended Plot**

Sludge Application	Year	Orchardgrass					Birdsfoot Trefoil				
		%N	%P	%K	%Ca	%Mg	%N	%P	%K	%Ca	%Mg
Mg ha⁻¹											
0	1977[a]										
	1978	0.92	0.18	1.51	0.36	0.23	1.03	0.24	1.93	0.61	0.26
	1979	1.17	0.22	2.41	0.30	0.22	2.59	0.14	1.74	1.82	0.40
	1980	1.11	0.24	1.86	0.32	0.20	2.11	0.17	1.92	1.02	0.23
	1981	1.22	0.18	1.62	0.68	0.22	3.32	0.17	1.89	1.52	0.29
	1989	1.67	0.17	1.82	0.42	0.30	2.31	0.17	1.71	0.92	0.22
184	1977	2.62	0.40	2.84	0.84	0.31	3.64	0.27	1.46	1.99	0.28
	1978	1.26	0.37	2.01	0.49	0.28	1.27	0.36	2.30	0.59	0.25
	1979	1.33	0.51	2.53	0.47	0.23	3.57	0.25	1.56	0.54	0.20
	1980	1.70	0.42	2.65	0.45	0.23	2.93	0.25	1.62	1.27	0.18
	1981	2.57	0.37	2.38	0.53	0.26	4.03	0.26	1.69	1.14	0.23
	1989	2.36	0.37	2.24	0.45	0.22	2.38	0.16	1.01	0.65	0.23

Source: Sopper and Seaker (1990).

[a] No plants available for sampling.

In August, 1989, the site was revisited and samples of vegetation, soils, and groundwater were collected to evaluate the long-term effects. Vegetation foliar samples were analyzed for nutrients and trace metals. For brevity, only the foliar analyses for the two dominating species, orchardgrass and crownvetch, will be discussed. The results of the soils and groundwater sample analyses are discussed in other sections of this book.

Foliar concentrations of macronutrients are given in Table 42. Nutrients (N and P) were all generally higher in the sludge-grown plants. Potassium and Ca were higher in the sludge-grown orchardgrass than in control plants. Potassium and Ca were only slightly lower in the sludge-grown birdsfoot trefoil plants than in the control plants. Foliar Mg concentrations were similar in both sludge-grown and control plants. Nutrient levels in the sludge-grown plants in 1989 were about the

Table 43. **Mean Foliar Concentrations (%) of Macronutrient Elements in Crownvetch on Sludge-Amended Plot**

Year	Crownvetch				
	% N	% P	% K	% Ca	% Mg
1977	3.36	0.34	1.64	2.63	0.42
1978	2.35	0.37	3.14	0.96	0.45
1979	3.35	0.22	1.29	1.68	0.25
1980	3.00	0.27	1.89	1.72	0.29
1981	3.78	0.31	1.89	1.25	0.23
1989	2.62	0.21	2.20	0.91	0.26

Source: Sopper and Seaker (1990).

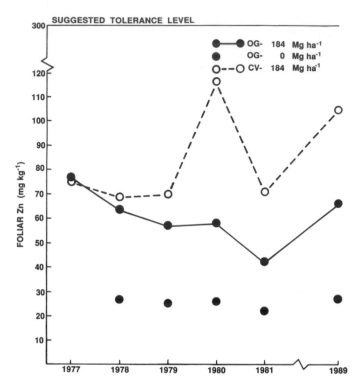

Figure 7.   Mean foliar concentration of Zn in orchardgrass and crownvetch collected from the control (0 Mg ha^{-1}) and sludge-amended (184 Mg ha^{-1}) plots.

same level as the first year when sludge was applied. There appears to be little depletion of nutrients from the site over the 12-year period. Birdsfoot trefoil data are given in Table 42 because no crownvetch plants were present on the control plot for comparison. Macronutrient concentrations in crownvetch on the sludge-amended plot are given in Table 43. Concentrations were quite similar to those of birdsfoot trefoil.

Foliar concentrations of Zn, Cu, Pb, Ni, Cd, Co, Cr, and B in orchardgrass and crownvetch are shown in Figures 7 to 14. Concentrations of Zn (Figure 7), Ni (Figure 10), and B (Figure 14) tended to be higher in crownvetch than in orchardgrass; whereas, concentrations of Cu (Figure 8) tended to be higher in orchardgrass. Concentration of Pb (Figure 9) and Cd (Figure 11) were variable and showed no distinct trends. In general, trace metal foliar concentrations tended to be highest the first year and then decrease over time. Except for Ni and Cr, foliar concentrations of trace metals in the sludge-grown orchardgrass plants were higher than in control plants. The 1989 values for Cu (Figure 8), Cd (Figure 11), Cr (Figure 13), and B (Figure 14) were quite similar to those of 1981. Foliar concentrations of Zn, Ni, Co, and Pb showed a slight increase from 1981 to 1989. Although sludge application appeared to increase some trace metal concentrations in the foliage, these increases were minimal and well below the suggested toler-

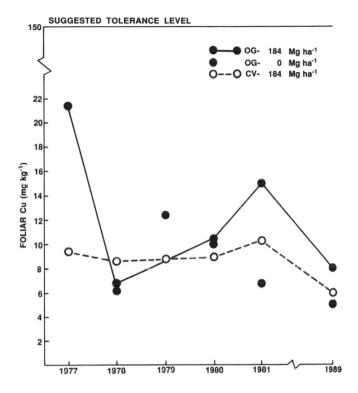

Figure 8.   Mean foliar concentration of Cu in orchardgrass and crownvetch collected from the control (0 Mg ha⁻¹) and sludge-amended (184 Mg ha⁻¹) plots.

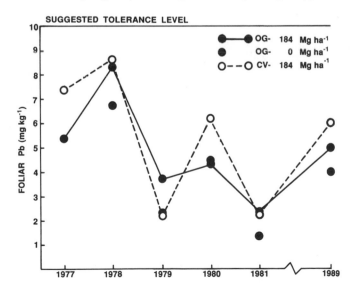

Figure 9.   Mean foliar concentration of Pb in orchardgrass and crownvetch collected from the control (0 Mg ha⁻¹) and sludge-amended (184 Mg ha⁻¹) plots.

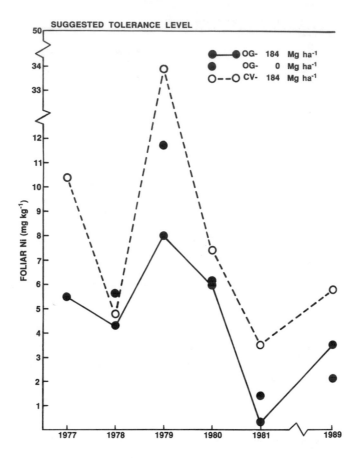

Figure 10. Mean foliar concentration of Ni in orchardgrass and crownvetch collected from the control (0 Mg ha⁻¹) and sludge-amended (184 Mg ha⁻¹) plots.

ance levels for agronomic crops (Melsted 1973). No phytotoxicity symptoms were observed during the study. The suggested tolerance levels are not phytotoxic levels but suggest foliar concentration levels at which decreases in growth may be expected.

In the second study, Sopper (1991a) resampled a deep coal mine refuse bank in 1990, 12 years after it had been amended with 80 and 108 Mg ha⁻¹ of dewatered municipal sewage sludge from the Scranton, PA wastewater treatment plant. The sludge and agricultural lime (11 Mg ha⁻¹) were applied to a 4 ha area and incorporated in May, 1978. Vegetation, soil, and groundwater samples were collected over a five-year period (1978 to 1982) and the results reported by Seaker and Sopper (1983). In August, 1990, the site was revisited and samples of vegetation and soil were collected to evaluate the long-term effects. The average concentrations of elements applied in the sludge are given in Table 44. The amounts of trace metals applied are given in Table 45 along with U.S. EPA and PDER recommendations.

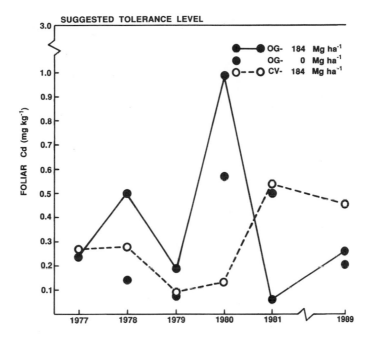

Figure 11. Mean foliar concentration of Cd in orchardgrass and crownvetch collected from the control (0 Mg ha^{-1}) and sludge-amended (184 Mg ha^{-1}) plots.

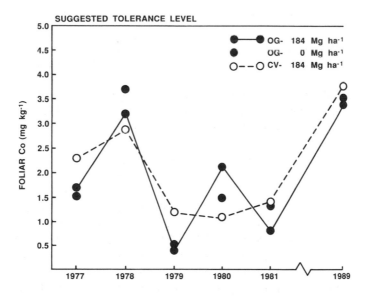

Figure 12. Mean foliar concentration of Co in orchardgrass and crownvetch collected from the control (0 Mg ha^{-1}) and sludge-amended (184 Mg ha^{-1}) plots.

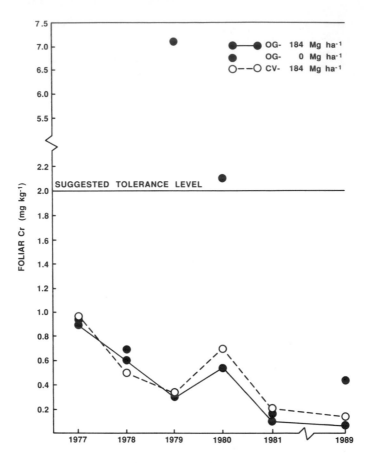

Figure 13.   Mean foliar concentration of Cr in orchardgrass and crownvetch collected from the control (0 Mg ha⁻¹) and sludge-amended (184 Mg ha⁻¹) plots.

Immediately after sludge application, the site was broadcast seeded with a mixture of K-31 tall fescue (28 kg ha⁻¹), Pennlate orchardgrass (28 kg ha⁻¹), Penngift crownvetch (17 kg ha⁻¹), and Empire birdsfoot trefoil (17 kg ha⁻¹), then mulched with straw and hay at 3.3 Mg ha⁻¹.

The site had a complete vegetative cover by August, 1978, two months after sludge application and has persisted over the 12-year period with no signs of deterioration. During the first two years the two grass species dominated the site, but by the third growing season the two legume species predominated. However, by 1990 the two grass species had disappeared and the dominating vegetative cover consisted primarily of crownvetch (90%) and birdsfoot trefoil (10%). A few volunteer tree species have also invaded the plot. The first year average height ranged from 35 to 47 cm, with the greatest growth being produced with the 108 Mg ha⁻¹ sludge application (Table 46). The greatest vegetation height increase occurred the second year after sludge application, then leveled off. Dry matter production significantly increased during the first three growing seasons and then decreased (Figure 15). Dry matter production values were higher than average hay

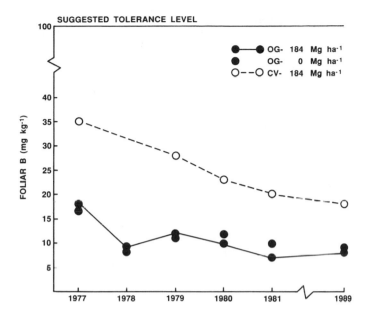

Figure 14. Mean foliar concentration of B in orchardgrass and crownvetch collected from the control (0 Mg ha^{-1}) and sludge-amended (184 Mg ha^{-1}) plots.

yield for conventional farmland in Lackawanna County (Pennsylvania Crop Reporting Service, 1980). In 1990 dry matter production was essentially the same as it was in 1982.

Foliar concentrations of macronutrients in crownvetch and birdsfoot trefoil over the 12-year period are given in Table 47. Concentrations of macronutrients remained relatively constant over the first five years after sludge application. After 12 years, only foliar concentrations of Ca and Mg show a substantial decrease. The data indicate that sufficient macronutrients were supplied in the sludge to maintain the vegetative cover for 12 years. There appears to be little depletion of the three major nutrients (N, P, K) from the site. Macronutrient values are generally in the intermediate range of these elements in various grasses and legumes as reported by Chapman (1966). Figures 16 to 24 show foliar concentrations of B and trace metals over the 12-year period. Concentrations of most trace metals showed definite decreases over the first five years (1978 to 1982). By 1990, 12 years after sludge application, concentrations of B, Mn, Cu, Co, and Ni remained low. However, concentrations of Pb, Zn, Cd, and Cr were higher in 1990 than 1982. This might be partially explained by the decrease in refuse pH. These results indicate that the sludge-borne metals become less available to plants due to binding by soil and organic matter. Also the increasing foliar biomass tends to dilute these metals in the plant tissue resulting in lower foliar concentrations. Concentrations of trace metals in the forage were well below the suggested tolerance levels for agronomic crops (Melsted, 1973) and no phytotoxicity symptoms were observed. The suggested tolerance levels are not phytotoxic levels but suggest foliar concentration levels at which decreases in growth may be expected.

Table 44.   Chemical Analysis of
Dewatered Sludge Applied
on Anthracite Refuse Bank

Constituent	Mean
pH	8.9
	mg kg^{-1}
Total P	4190
NO_3–N	137
NH_4–N	3271
Org–N	12215
Ca	79257
Mg	4724
Na	714
K	1292
Al	6550
Mn	1427
Fe	31903
Co	8
Zn	776
Cu	865
Pb	613
Cr	202
Ni	57
Cd	16.4
Hg	1.3
	%
Total P	0.42
Total N	1.56
Solids	28.83

Source: Sopper (1991a).

Table 45.   Comparison of Trace Metal Loadings with EPA and PDER
Recommendations

	Sludge Application Rate (Mg ha^{-1})		Recommendations	
Element	80	108	EPA (CEC 5–15)[a]	PDER
	kg ha^{-1}			
Cd	1.2	1.7	11	3
Cu	67	92	280	112
Cr	16	21	NR[b]	112
Pb	49	67	800	112
Hg	0.12	0.17	NR[b]	0.6
Ni	4.4	5.9	280	22
Zn	64	86	560	224

Source: Sopper and Seaker (1984b).

[a]   Average CEC of site ranged from 7.8 to 10.4 meq 100 g^{-1}. From U.S. EPA (1983).
[b]   No recommendation given by EPA.

Table 46.   Average Vegetation Height Growth

Sludge Application Mg ha⁻¹	Height (cm)				
	1978	**1979**	**1980**	**1981**	**1990**
80	35	82	46	36	55
108	47	81	53	38	57

Source: Sopper (1991a).

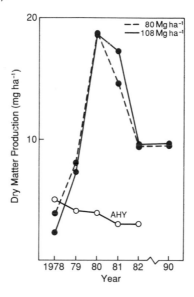

Figure 15.   Dry matter production on the refuse bank reclaimed with dewatered sludge compared to average hay yields in Lackawanna County.

Table 47.   Mean Foliar Concentrations of Macro-nutrients in Crownvetch and Birdsfoot Trefoil Collected from 108 Mg ha⁻¹ Sludge Application Plot

Year	%N	%P	%K	%Ca	%Mg
			Crownvetch		
1978	3.09	0.20	1.63	1.86	0.25
1979	3.54	0.29	1.62	2.62	0.69
1980	3.23	0.20	1.73	1.89	0.56
1981	3.57	0.20	1.37	2.04	0.65
1982	3.46	0.21	1.41	1.98	0.63
1990	4.70	0.36	1.73	1.41	0.61
			Birdsfoot Trefoil		
1978	3.1	0.17	1.60	1.60	0.30
1979	3.5	0.25	1.38	1.86	0.56
1980	2.6	0.20	1.43	1.69	0.41
1981	3.3	0.18	1.12	1.77	0.48
1982	3.9	0.20	1.41	2.20	0.50
1990	3.1	0.21	1.55	0.59	0.31

Source: Sopper (1991a).

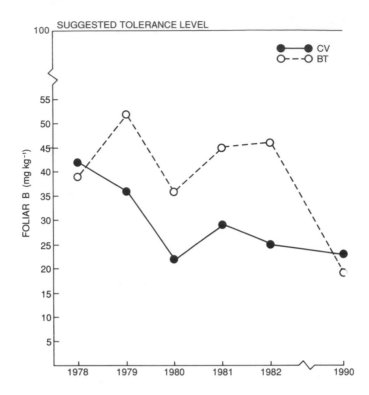

Figure 16. Mean foliar concentration of B in the crownvetch and birdsfoot trefoil collected from the 108 Mg ha⁻¹ sludge application plot.

Samples of panic grass (*Panicum dichotomiflorum* Michx.) that volunteered on the sludge treated and control areas, were also analyzed for trace metals (Table 48). Results indicate that there was little difference in metal concentrations on the sludge treated plots compared to the untreated area during the first two years. In fact, concentrations of Cr, Pb, and Ni were higher in the panic grass from the control area than in the panic grass grown with sludge. By 1990, foliar concentrations of Cu, Zn, Pb, and Ni remained the same or decreased substantially. Foliar concentrations of Cd, Cr, and Cu increased substantially, but the values were still well below the suggested tolerance levels. Foliar concentrations of all trace metals were similar for plants growing on both the control area and the sludge amended plots.

Another long-term study has been reported by Palazzo and Reynolds (1991). They evaluated metal foliar concentrations in plants growing on a dredge disposal area in Delaware which received 100 Mg ha⁻¹ of anaerobically digested and dewatered sludge and 23 Mg ha⁻¹ of lime. Plant samples were collected at 2, 4, and 16 years after sludge application. Foliar metal concentrations in three grass species (tall fescue, red fescue, and Kentucky bluegrass) are given in Table 49. Foliar concentrations of Cu, Zn, Ni, and Cd all decreased substantially over the 16-year period. Only foliar concentrations of Cr and Pb were higher in year 16 than in year 2. However, only

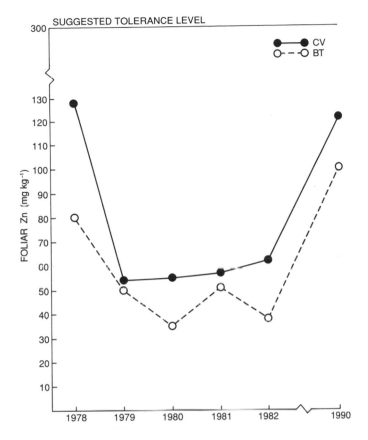

Figure 17. Mean foliar concentration of Zn in the crownvetch and birdsfoot trefoil collected from the 108 Mg ha⁻¹ sludge application plot.

foliar concentrations of Cr (5.6 mg kg⁻¹) exceeded the suggested tolerance level (2.0 mg kg⁻¹) cited by Melstead (1973) which might cause a decrease in plant growth.

The Philadelphia Water Department has the largest ongoing sludge application program for land reclamation in Pennsylvania. Approximately 1134 ha of surface-mined land were reclaimed between 1978 and 1987 using Philadelphia sludge as the minespoil amendment. Carrello (1990) and Seaker (1991) reported a ten-year summary of the environmental monitoring on some of these sites which included sampling of vegetation, soils, soil percolate water, and groundwater. The sludge applied on all 49 sites was Philadelphia mix. This mix consists of anaerobically digested and dewatered sludge cake (20% solids) mixed with an equal volume of composted sludge. The design sludge application rate was approximately 134 Mg ha⁻¹. Actual application rate on each site varied slightly between years depending on sludge trace metal concentrations. All sites were limed to raise spoil pH to a minimum of 6.5. After lime and sludge were incorporated, all sites were seeded with a mixture of tall fescue (22 kg ha⁻¹), orchardgrass (22 kg ha⁻¹), and birdsfoot trefoil (11 kg ha⁻¹).

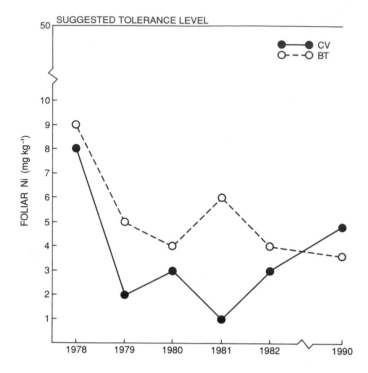

Figure 18.   Mean foliar concentration of Ni in the crownvetch and birdsfoot trefoil collected from the 108 Mg ha^{-1} sludge application plot.

Average dry matter production on sites sampled one to five years after sludge application ranged from 2.7 to 10.2 Mg ha^{-1} compared to average hay yields of 2.2 Mg ha^{-1} from conventional farm fields in the county.

Foliar concentrations of Cu, Cd, Zn, and Pb were very consistent among sites sampled one to seven years after sludge application (Table 50). Copper concentrations ranged from 4 to 30 mg kg^{-1}, which is comparable to the range of 11 to 31 mg kg^{-1} for plants from fertilizer-amended mine sites, and well below the suggested crop tolerance level of 150 cited by Melsted (1973). Eighty-seven percent of the foliar samples had Cu levels within the same range as the maximum levels found in selected food crops from U.S. agricultural areas. Ninety-six percent of the samples had Cu levels below the tolerance limit for animal diets. Foliar concentrations of Cd ranged from 0.01 to 1.75 mg kg^{-1} and Zn ranged from 16 to 182 mg kg^{-1}, which were well below the crop tolerance levels. All Cd levels were within the same range as maximum levels found in selected food crops from U.S. agriculture areas. The majority of the Zn samples were within the same range as maximum levels found in selected food crops from U.S. agricultural areas and all samples had levels below the animal tolerance level. Eighty-one percent of the samples were lower in Zn than the minimum requirement for dairy rations of 70 mg kg^{-1}. Foliar Pb concentrations ranged from 0.01 to 13 mg kg^{-1}. Ninety-six percent of the samples were below the crop tolerance level, and all were below the animal dietary tolerance level. However, Pb concentrations were higher than the maximum levels found in selected food crops from U.S. agricultural areas.

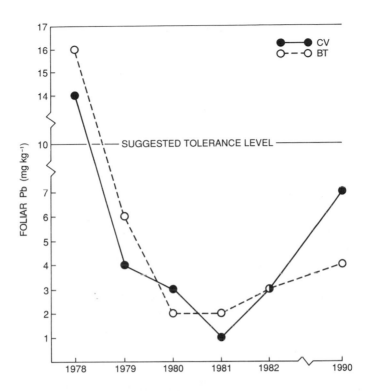

Figure 19. Mean foliar concentration of Pb in the crownvetch and birdsfoot trefoil collected from the 108 Mg ha^{-1} sludge application plot.

Total metal concentrations in the plow layer of the minespoil on sites sampled one to eight years after sludge application are shown in Table 51. Copper, Cd, Zn, and Pb were higher on the sludge-amended sites than on mine sites amended with chemical fertilizer, as would be expected. Copper and Pb, however, were consistently within the ranges of these metals found in agricultural soils in the United States. For Cd, 14 of the 18 sites sampled were within the range for U.S. agricultural soils, while the other four were only slightly above (2.3 vs 2.5 to 3.68 mg kg^{-1}). For Zn, all but one site were within the range for agricultural soils. The data show that metals applied in the sludge may significantly increase metal levels in the plow layer, but the resulting levels were acceptable for agricultural production.

Ranges of nitrate-N and metal concentrations in soil percolate water collected at a depth of 90 cm monthly and compiled on a quarterly basis are given in Table 52. As would be expected, the nitrate-N concentrations were increased above background levels on some sites for the first three quarters prior to vegetation establishment, and then subsequently declined. Concentrations of Cd, Cu, Zn, and Pb; however, showed a decrease over time compared to background levels. For six quarters, metal concentrations in the percolate water remained low. The percentage of samples meeting drinking water standards are given in Table 53. The percentage of samples meeting drinking water standards for metals was similar before and up to 18 months after sludge was applied.

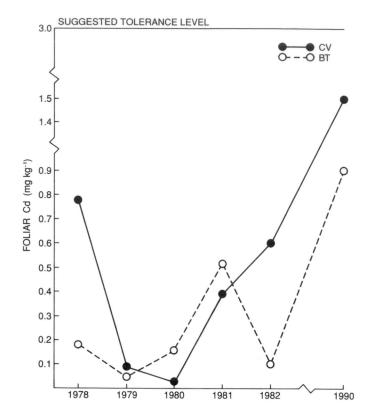

Figure 20.   Mean foliar concentration of Cd in the crownvetch and birdsfoot trefoil collected from the 108 Mg ha⁻¹ sludge application plot.

Depth to the water table on the sites ranged from 1 to 34 m. The majority of the groundwater monitoring wells ranged from 9 to 25 m. Concentrations of NO_3-N and trace metals in groundwater sampled monthly and compiled on a quarterly basis are given in Table 54. Groundwater samples analyzed up to 18 months after sludge application had nitrate and metal concentrations generally in the same range as the background samples, indicating that the sludge application had no detectable effect on water quality during that period. The initial increased NO_3-N levels in the percolate water did not affect groundwater NO_3-N concentrations, probably due to dilution effects. The percentage of groundwater samples meeting U.S. EPA drinking water standards are given in Table 55. The percentages of groundwater samples meeting drinking water standards were similar before, and up to 18 months after, sludge application.

Carrello (1990) concluded that Philadelphia mine mix sludge has proven to be beneficial in the reclamation of surface-mined land, and has been conclusively shown to cause no adverse effects on groundwater quality, soils, or vegetation.

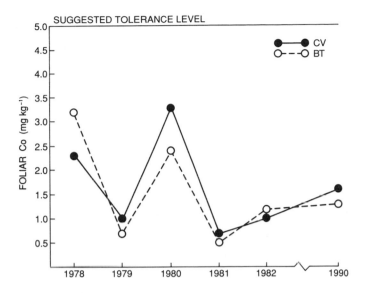

Figure 21.   Mean foliar concentration of Co in the crownvetch and birdsfoot trefoil collected from the 108 Mg ha⁻¹ sludge application plot.

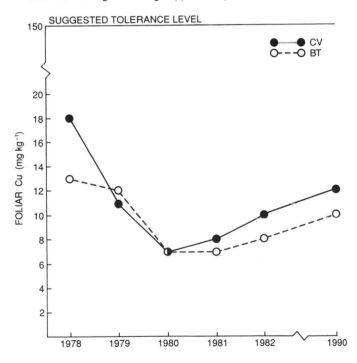

Figure 22.   Mean foliar concentration of Cu in the crownvetch and birdsfoot trefoil collected from the 108 Mg ha⁻¹ sludge application plot.

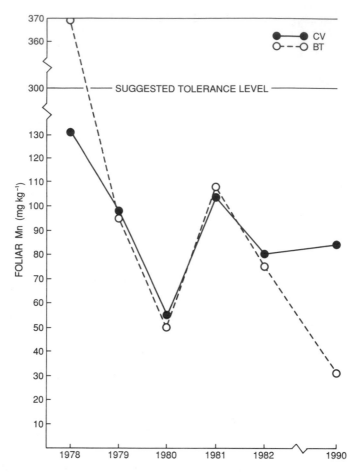

Figure 23. Mean foliar concentration of Mn in the crownvetch and birdsfoot trefoil collected from the 108 Mg ha⁻¹ sludge application plot.

In 1990, the U.S. Environmental Protection Agency, Region III, reported the results of a study which reevaluated six sites which had been reclaimed using Philadelphia mix (U.S. EPA, Region III and Gannett Fleming Environmental Engineers, Inc., 1990). These six sites were selected from a total of 74 reclamation sites using municipal sludge within the Region's jurisdiction. The study included the collection of vegetation, soil, soil percolate water, and groundwater samples. Results were evaluated in comparison to Federal and/or state regulatory standards for drinking water, plant tolerance levels, animal consumption rates, soil concentration limits, and agricultural irrigation water limits in order to identify potential environmental degradation and possible public health threats associated with sludge reclamation projects. Six "sister" sites or control sites reclaimed by conventional methods were also evaluated. The conclusions of the study can be summarized as follows:

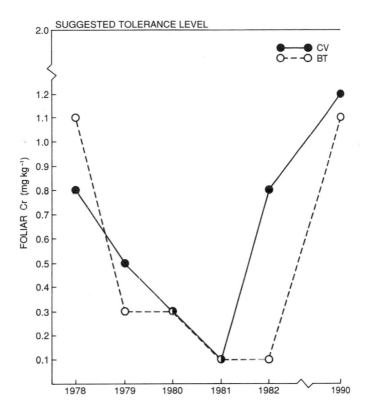

Figure 24. Mean foliar concentration of Cr in the crownvetch and birdsfoot trefoil collected from the 108 Mg ha⁻¹ sludge application plot.

1. Metals, while detected in soils, vegetation and/or soil percolate water, were generally within applicable standards and tolerance limits on both sludge amended and conventional reclaimed sites.
2. Metals in vegetative tissues were generally found to be below plant tolerance levels for phytotoxicity which could result in vegetative stress, reduced crop yield, or plant death. No evidence of vegetative stress or phytotoxicity was apparent on any site, even where plant tolerance levels were exceeded.
3. Metals in vegetative tissues were generally found to be below the most conservative tolerance levels for vegetative consumption by wild and domestic animals.
4. Sludge application at levels designed for reclamation does not appear to elevate metals within the soil to levels that would affect plants and soil biota.
5. Organic compounds were not identified as a problem at either sludge amended or conventionally reclaimed sites. With the exception of DDE, organic compounds were not detected at any site. Levels of DDE were below plant and soil biota toxicity limits and DDE was not detected in any groundwater samples.
6. After reclamation with sludge, soil percolate water and groundwater generally met drinking water standards, except for pH which was lower than the prescribed range, a situation typical of acid mine drainage areas throughout the region.

Table 48.   Mean Foliar Concentrations of Trace Metals in Panic Grass Collected from Control Area and Sludge Application Plots

Sludge Application Rate (Mg ha⁻¹)	Year	Concentrations (mg kg⁻¹)						
		Cu	Zn	Cr	Pb	Co	Cd	Ni
0	1978	9	41	0.9	3.7	0.8	0.11	4.3
80		15	40	0.5	2.3	1.3	0.15	2.6
108		10	59	0.7	3.0	1.4	0.11	3.3
0	1979	11	41	0.4	5.0	0.2	0.07	2.7
80		16	34	0.2	1.0	0.3	0.05	1.9
108		20	44	0.3	2.3	0.3	0.06	1.8
0	1990	2	29	0.8	1.0	0.7	0.84	2.0
80		4	27	0.9	1.0	0.6	0.84	1.9
108		5	29	1.0	1.0	0.7	0.84	2.0
Suggested tolerance levels[a]		150	300	2	10	5	3	50

Source: Sopper (1991a).

[a] Melsted, 1973.

Trees. Only a limited number of studies have been done with metal uptake by trees. Trees are beneficial for sludge-amended sites because they are not a significant food source, and because of their large biomass, they tend to retain heavy metals on the site through biotic storage in plant parts not grazed by herbivores, thereby limiting their entry into food chains (Roth et al., 1982; Morin, 1981). Two studies looked at accumulation of metals by component tree parts. Patterns of accumulation by roots, stems, and leaves varied with the metal and the tree

Table 49.   Mean Plant Foliar Trace Metal Concentrations 2, 4, and 16 Years After Sewage Sludge Application

Element	Year	Concentrations (mg kg⁻¹)	
		Mean	Range
Cu	2	8.9	8.2–9.6
	4	9.0	8.0–10.0
	16	5.6	4.9–6.1
Zn	2	82.9	64.0–107.0
	4	165.3	74.0–291.0
	16	9.3	6.7–13.0
Cr	2	1.3	0.9–2.0
	4	10.3	10.0–11.0
	16	5.6	4.9–6.1
Ni	2	14.4	10.7–18.6
	4	2.3	2.0–3.0
	16	2.3	2.3
Pb	2	0.2	0.1–0.2
	4	5.7	5.0–6.0
	16	2.0	2.0–4.0
Cd	2	0.6	0.5–1.0
	4	0.6	0.5–0.7
	16	<0.2	<0.2

Source: Palazzo and Reynolds (1991).

Table 50. Metal Concentrations (mg kg⁻¹) in Forage Sampled 1 to 7 Years After Sludge Application, Compared to Tolerance Levels for Crops and Animal Diets, and to Concentrations in Crops Grown on U.S. Farms

Year	mg kg⁻¹			
	Cu	Cd	Zn	Pb
1–2	4–30	0.01–1.2 (2.7)	16–182 (231)	0.01–13
4	6–27 (59)	0.5–0.8	34–79 (127)	2–11
5	8–15	0.5–0.75	37–94	3–4
6	12–14	0.5–1.75	47–94	2–4
7	8–29	0.5–0.8	48–84	2–6
Crop tolerance level[a]	150	3	300	10
Animal tolerance level[b]	25–800	0.5	300–1000	30
Minimum in dairy rations[c]	11	—	70	—
Fertilizer-amended minesites[d]	11–31	0.3–0.5	34–87	3–7
Maximum in U.S. farm crops[e]				
Spinach	20	1.90	200	2.30
Lettuce	18	3.85	110	1.70
Potato	14	1.00	35	2.20
Wheat grain	10	0.22	76	0.77

Note: Ranges are given with outliers in parentheses.

Source: Adapted from Carrello (1990) and Seaker (1991).

[a] Melsted (1973).
[b] National Research Council (1980).
[c] Pennsylvania State University (1970).
[d] U.S. EPA and Gannett-Fleming Environmental Engineers (1990).
[e] Wolnick et al. (1983a, 1983b, 1985)

Table 51. Ranges in Total Metal Concentrations in Plow Layer from Sites Sampled 1 to 8 Years After Sludge Application, Compared to Unamended Minespoil, Fertilizer-amended Minespoil, and U.S. Soils

Year	(mg kg⁻¹)				Number of Sites
	Cd	Cu	Zn	Pb	
0[a]	0.02–1.5	12–48	22–110	17–38	5
1	0.12–1.05	22–67	57–95	19–66	3
2	0.20–1	31–39	81–124	46–65	2
4	2.5–2.8	87–101	225–243	55–61	2
5	1.2	45	118	38	1
6	0.95–2.5	53–124	142–339	34–77	5
7	0.78–3.68	47–191	123–511	36–128	4
8	1.3	65	198	70	1
Fertilizer amended Minespoil[b]	1	32–61	71–158	23–33	5
U.S. Soils[c]	0.01–2.3	0.3–735	1.5–402	0.2–4109	

Source: Carrello (1990).

[a] Background samples.
[b] U.S. EPA and Gannett-Fleming Environmental Engineers (1990).
[c] Sommers et al. (1987).

Table 52.   Ranges in NO₃–N and Metal Concentrations in Spoil Percolate Water Sampled Before Sludge Application (0), and for 6 Quarters Following Sludge Application[a]

Quarter	n	NO₃–N	Cd	Cu	Zn	Pb
				mg L^1		
0	40	0.15–48	<0.001–0.055 (0.177)	<0.01–4.5	0.02–44	<0.01–0.36
1	45	0.4 –101	<0.001–0.070 (0.225)	<0.01–1.8 (4.8)	0.04–29 (49)	<0.01–0.25
2	44	0.14–63	<0.001–0.036	<0.01–0.80 (1.65)	0.01-4 (22)	<0.01–0.23
3	43	0.01–92	<0.001–0.020	<0.01–0.86	0.01–8	<0.01–0.18
4	41	0.28–52	<0.001–0.016	<0.01–0.86	0.01–6	<0.01–0.14 (0.34)
5	18	0.25–41 (93)	<0.001–0.031	0.01–0.55 (1.12)	0.02–5 (11)	<0.01–0.15
6	8	0.18–1.2 (63)	<0.001–0.037	0.01–0.22 (2.17)	0.04–2 (17)	<0.01–0.10

Note: Outliners are shown in parentheses; N = number of sites.

Source: Carrello (1990).

[a]   Background samples.

species. Eastern white pine, silver maple, and green ash, for example, tended to accumulate Cu, Zn, and Cd in their roots as opposed to foliage and stems (Svoboda et al., 1979). Roots and leaves were higher accumulators than stems (Roth et al., 1982; Morin, 1981). In a study using a high-metal sludge, metals were sometimes lower in the leaves of sludge-grown trees than in the fertilizer controls, except Cd in hardwoods, which was consistently higher when sludge was used (Schneider et al., 1981). Roth and others (1982) observed lower metal concentrations the third growing season after sludge application compared to the first, but some third-year concentrations exceeded those considered to be toxic. For example, the maximum Cd concentration of combined tree parts was 950 mg kg^{-1}. The ability of the tree to tolerate such high levels of metal may be due to a tie-up of metal ions in plant tissues in forms not readily movable throughout the plant.

Table 53.   Percentage of Percolate Water Samples Meeting U.S. EPA Drinking Water Standards Before and After Sludge Application

	U.S. EPA[a] D.W.S. (mg L^{-1})	% Within D.W.S. Before Sludge	% Within D.W.S. After Sludge
NO₃–N	10	81	67
Cd	0.010	86	91
Cu	1	95	98
Zn	5	89	96
Pb	0.05	81	71

Source: Carrello (1990).

[a]   U.S. EPA (1979).

Table 54. Ranges in NO_3–N and Metal Concentrations in Groundwater from 42 Sites Sampled Prior to Sludge Application (0), and for 6 Quarters Following Sludge Application[a]

Quarter	n			Concentrations (mg L^{-1})		
		NO_3–N	Cd	Cu	Zn	Pb
0	40	<0.1–9	<0.001–0.04	<0.01–1.28	<0.01–11	<0.01-0.22
		(25)	(0.08)	(2.27)	(26)	(0.66)
1	42	<0.1–12	<0.001–0.02	<0.01–1.66	0.01–14	<0.01–0.31
			(0.12)	(3.45/4.19)	(89)	(0.42)
2	42	<0.1–5.2	<0.001–0.03	<0.01–1.54	<0.01–12	<0.01–0.24
		(7)			(35)	(0.58)
3	41	<0.1–13	<0.001–0.02	<0.01–1.13	0.01–11	<0.01–0.27
		(16)		(1.98)		
4	38	<0.1–10	<0.001–0.02	<0.01–1.08	0.02–9	<0.01–0.21
		(21)		(1.92)		
5	23	<0.1–4	<0.001–0.04	<0.01–0.73	0.01–11	<0.01–0.16
		(8)	(0.35)	(1.75/2.82)		
6	12	<0.1–4	<0.001–0.02	<0.01–0.68	0.23–12	<0.01–0.21
		(6)		(2.3/4.1)		

Note: Outliners are shown in parentheses; N = number of sites.

Source: Carrello (1990).

[a] Background samples.

In Colorado, Sabey et al. (1990) reported that growing fourwing saltbush and mountain big sagebush in copper mine spoil treated with 60 Mg ha^{-1} of air-dried municipal sewage sludge increased leaf tissue concentrations of Zn, Fe, Mn, and Pb; however, none of the values were in the extremely high level reported by Chapman (1982). Concentrations of Cu and Cd in shrub leaves were greatly above the normal range for plant tissue; however, only Cd approached the suggested tolerance level reported by Melsted (1973).

Kormanik and Schultz (1985) reported on the use of digested sewage sludge applications on borrow pits in South Carolina to establish tall fescue and sweetgum seedlings. Sludge application rates were 0, 17, 34, and 68 Mg ha^{-1}. Results showed that the current rate of sweetgum growth after sludge and subsoiling was comparable to that of sweetgum established in plantations on routine reforestation sites. After

Table 55. Percentage of Groundwater Samples Meeting U.S. EPA Drinking Water Standards Before and After Sludge Application

	U.S. EPA[a] D.W.S. (mg L^{-1})	% Within D.W.S. before sludge	% Within D.W.S. after sludge
NO_3-N	10	98	98
Cd	0.010	94	95
Cu	1	99	97
Zn	5	93	92
Pb	0.05	79	65

Source: Carrello (1990).

[a] U.S. Environmental Protection Agency (1979).

Table 56.   Effects of Sludge-Fly Ash Amendments on Selected Trace Metal Concentrations in the Foliage of Tree Seedlings

Tree species	Amendment Mixture	Concentrations (mg kg⁻¹)						
		Mn	Cu	B	Zn	Pb	Cd	Ni
White pine	1S:1FA	947	14	56	700	40	14	4
	2S:1FA	791	13	47	840	31	6	4
	3S:1FA	732	13	33	604	49	14	4
Larch	1S:1FA	103	7	66	408	34	6	3
	2S:1FA	170	8	84	402	33	5	3
	3S:1FA	159	7	71	450	42	4	3
Black locust	1S:1FA	103	12	104	870	88	11	6
	2S:1FA	168	12	81	430	59	7	6
	3S:1FA	121	12	72	470	47	12	6
Alder	1S:1FA	154	25	57	1258	140	5	7
	2S:1FA	441	20	58	807	104	4	7
	3S:1FA	164	21	56	2070	76	3	10
Black cherry	1S:1FA	121	12	29	1033	99	7	5
	2S:1FA	215	11	34	887	84	6	3
	3S:1FA	170	9	28	490	53	4	4
Hybrid poplar	1S:1FA	106	9	64	1605	34	32	3
	2S:1FA	179	10	34	1650	70	27	4
	3S:1FA	89	10	47	4000	28	66	4
Suggested tolerance level[a]		300	150	100	300	10	3	50

Source: Sopper (1989c).

[a]   Melsted (1973).

two years, tree heights were significantly greater on the sludge-amended plots than on nonamended plots. Root-collar diameter growth followed the same trend. Tree survival for nonamended plots and for the 17, 34, and 68 Mg ha⁻¹ sludge treatments were 76, 54, 52, and 42%, respectively. Lower survival on the sludge plots was attributed to competition by the heavier grass biomass.

In Pennsylvania, Sopper (1989a) reported on the uptake of trace metals by tree seedlings planted on a contaminated zinc smelter site amended with mixtures of sludge and fly ash. The tree species evaluated and foliar trace metal concentrations are given in Table 56. Foliar concentrations of Cu, Ni, Cr, and Co were under the suggested tolerance level for all tree species on all amendments. Similarly, foliar B concentrations were all below the suggested tolerance level for all tree species on all amendments except for black locust on the 1S:1FA mixture. Foliar Mn concentrations were all below the suggested tolerance level except for white pine on all three amendments and alder on the 2S:1FA mixture. Foliar Pb concentrations exceeded the suggested tolerance level in all species on all amendments. The highest foliar Pb concentration was 140 mg kg⁻¹ in alder. Foliar Cd concentrations equaled or exceeded the suggested tolerance level in all species on all amendments. Hybrid poplar had the highest Zn and Cd concentrations. Alder and black cherry had the lowest Cd concentrations and larch had the lowest Zn concentrations. In general, within a species there was little difference in foliar trace metal

Table 57.   Amounts of Trace Metals in 0 to 15 cm Depth of Surface Soil on Palmerton Plots Treated With Sludge-Fly Ash Mixture (2S:1FA) and Vegetation Foliar Concentrations

	Zn	Cu	Pb	Ni	Cd	Mn
	kg ha^{-1}					
Amount of element in 0-15 cm soil depth	1210	8	1	1	47	8
Amount added in sludge-Fly ash mixture	97	74	17	16	1	464
Total trace metals	1307	81	18	17	48	472
Normal levels in soils[a]	40-150	4-80	40-50	40-50	1-4	40-200

Vegetation Foliar Concentration

	mg kg^{-1}					
Trees						
White pine	840	13	31	4	6	791
Austrian pine	239	8	17	2	3	395
Virginia pine	378	9	20	5	8	383
Red pine	174	5	22	2	1	928
Japanese parch	402	8	33	3	5	170
Black pocust	430	12	59	6	7	168
Arnot bristly locust	600	17	84	7	7	283
European alder	807	20	104	7	4	441
Black cherry	887	11	84	3	6	215
Sugar maple	940	17	200	3	12	780
Hybrid poplar	1650	10	70	4	27	179
Grasses						
Blackwell switchgrass	715	18	83	3	5	189
Cave-in-rock switchgrass	961	25	113	3	8	135
Niagara big bluestem	1089	22	118	2	7	98
Tall fescue	950	24	69	7	20	379
Ryegrass	1396	36	51	20	17	632
Oahe intermediate wheatgrass	470	21	38	3	8	168
Orchardgrass	1582	26	98	9	17	125
Legumes						
Lathco flatpea	2082	45	87	12	17	694
Birdsfoot trefoil	1121	36	59	12	26	250
Crownvetch	—[b]	—[b]	—[b]	—[b]	—[b]	—[b]
Tolerance level[c]	300	150	10	50	3	300

[a] Baker and Chesnin, 1974.
[b] No surviving plants.
[c] Melsted, 1973.

concentrations among the three amendments. No phytotoxicity symptoms were observed on any species.

Results from the Palmerton project provide some interesting results related to trace metal uptake by vegetation growing on the highly contaminated soil amended with sludge and fly ash and vegetation growth responses. A summary of these results are given in Table 57. Considerable amounts of trace metals were present in the soil and sludge-fly ash amendment for vegetation uptake. The amounts of trace metals shown in Table 57 are for extractable metals available for plant

Table 58.   Average Concentrations of Trace Metals in the Foliage of the Tree Species

Species	Plot	Concentrations (mg kg⁻¹)						
		Mn	Cu	B	Zn	Pb	Cd	Ni
Catalpa	C	92	10	16	32	2	0.13	6.7[b]
	2S:1FA	40	9	30	21	4[b]	0.05	3.2
Alder	C	299[a]	13	13	40	4	0.15	8.2
	2S:1FA	234	27[a]	21[a]	58[b]	6	0.35	8.6
Tree of heaven	C	245[c]	8	27	24	4	0.07	4.6
	2S:1FA	135	12[b]	61	34[a]	6	0.07	4.0
Red Oak	C	398	33	8	49	7	0.35	2.7
	2S:1FA	452[a]	43[b]	11	56[a]	9	0.45	3.4
Black locust	C	34	11	12	20	2	0.11	7.1
	2S:1FA	61	14[a]	32[a]	36[a]	4[a]	0.03	12.2
Larch	C	212	3	14	14	2	0.26	4.9
	2S:1FA	244[a]	9[a]	36[a]	22[a]	4[a]	1.05[a]	6.8
Suggested tolerance level[d]		300	150	100	300	10	3	50

Source: Sopper (1991b).

[a,b,c] Significant effect at P <0.05, 0.01, and 0.001, respectively.
[d] Melsted (1973).

uptake. Zinc, Cd, and Mn were very high considering what is normal for uncontaminated soils (Baker and Chesnin, 1975). Zinc was over twofold and Cd over fourfold higher than current U.S. EPA recommendations for maximum lifetime loadings per hectare for these elements. In all tree species, foliar concentrations of Cu and Ni were below the suggested tolerance level. All tree species had foliar concentrations of Pb over the suggested tolerance level. Highest Pb foliar concentration (200 mg kg⁻¹) was found in sugar maple. This high concentration, along with Mn toxicity, may have contributed to the poor vigor and survival of this species. All tree species foliar concentrations of Zn exceeded tolerance levels, except for red pine and Austrian pine. Hybrid poplar had the highest foliar Zn concentration (1650 mg kg⁻¹) yet exhibited one of the most vigorous growth responses. Cadmium foliar concentrations in all species, except red pine, also exceeded the tolerance level. Manganese foliar concentrations exceeded the tolerance level in all species, except Japanese larch, Arnot bristly locust, black cherry, and hybrid poplar. Manganese foliar concentrations were very high in sugar maple seedlings (780 mg kg⁻¹) and many seedlings exhibited Mn toxicity symptoms—leaf margin, chlorosis, necrotic spots on the leaves, and leaf puckering. Jones (1972) reported the following ranges of Mn in native leaf tissue for most agricultural crops: deficient, <20 mg kg⁻¹; sufficient, 20 to 500 mg kg⁻¹; and toxic, >500 mg kg⁻¹. Foliar Mn concentrations in white pine, red pine, and sugar maple all exceeded 500 mg kg⁻¹. No phytotoxicity symptoms were observed on the pine species; however, they did exhibit poor growth and vigor.

Similarly to the tree species, foliar concentrations of Cu and Ni in the grass and legume species were all below the tolerance levels. Foliar concentrations of Zn, Pb, and Cd in the herbaceous species were all higher than the tolerance levels.

Foliar Mn concentrations were all lower than the tolerance level in all herbaceous species, except in tall fescue, ryegrass, and Lathco flatpea. There were no surviving plants of crownvetch, possibly due to a combination of phytotoxic effects. Foliar concentrations in crownvetch of Mn, Zn, Pb, and Cd were exceedingly high, and exceeded tolerance levels in the one treatment plot (1S:1FA) were a few plants survived.

Except for the few instances discussed above, the growth responses of most of the tree species and grass and legume species was exceptionally good, despite the high foliar concentrations of many of the trace metals.

More recently, Sopper (1991b) reported on the uptake of trace metals by tree seedlings planted on a steep coal refuse bank amended with a mixture of sludge and fly ash in Pennsylvania. Sludge application was equivalent to 65 Mg ha^{-1}.

Average concentrations of trace metals in the foliage of the tree species in 1990 are given in Table 58. Foliar trace metal concentrations were generally higher in the tree seedlings growing on the sludge-fly ash amended plots. All foliar trace metal concentrations in all species, except red oak, were below the suggested tolerance level for agronomic crops cited by Melsted (1973). The suggested tolerance levels are not phytotoxic levels but suggest foliar concentration levels at which decreases in growth may be expected. Red oak foliar Mn concentrations exceeded the suggested tolerance level in tree seedlings growing in both the control and amended plots. Symptoms of Mn toxicity — leaf margin chlorosis, necrotic spots on the leaves, and leaf puckering — were observed on many of the red oak seedlings. The poor height growth of this species might also be due to Mn toxicity. All species, except catalpa, had significantly higher foliar Cu and Zn concentration on the sludge-fly ash amended plots. Both black locust and larch had significantly higher foliar concentrations of B and Pb on the amended plots. Larch was the only species which had a significantly higher foliar Cd concentration on the amended plots.

Bayes, Taylor, and Moffat (1990) reported on the application of stored liquid undigested sludge on a degraded heathland soil in northern Scotland (Ardoss Forest site). Sludge application rates were 364 and 728 m^3 ha^{-1}. After the sludge was plowed, the site was planted with sitka spruce. After five years, the sitka spruce were 185 cm in height on the 364 m^3 ha^{-1} sludge site and 168 cm in height on the 728 m^3 ha^{-1} sludge site, in comparison to 125 cm for control trees.

Liquid digested sludge was also applied on a former open cast coal mine site (Clydesdale) at rates of 94 and 195 m^3 ha^{-1}. During the first year, the sludge applications increased sitka spruce needle weights per tree by 11 and 27% for the two sludge application rates, respectively. Tree annual height increments were also increased by 197 and 220% by the two sludge application rates, respectively, relative to control trees. Dewatered digested sludge was also applied at rates of 106 and 194 Mg ha^{-1} on former opencast coal mine sites at Clydesdale. The site which originally had a poor sward density quickly developed a dense sward cover, reducing surface runoff and erosion.

Arnot, McNeill, and Wallis (1990) also reported on the application of a liquid digested sludge at rates of 114 and 159 m^3 ha^{-1} to a young sitka spruce plantation on a former outcast coal mine at Clydesdale, Scotland. Total annual height growth increment of the sitka spruce trees was 29 and 44 cm on the two sludge-treated sites, respectively, in comparison to 20 cm for the control trees.

Field Crops. Although field crops are not often grown on reclaimed sites, some studies have been made to assess metal uptake where sludge is used as an amendment. Hundreds of hectares in Fulton Co., Illinois, that were once strip mined for coal were returned to corn production after the soil was reconditioned with sludge from Chicago. Cadmium loadings as high as 135 kg ha^{-1} resulted in some fields, and Cd concentrations were increased in corn grain, with maximum values in 1979 of 0.46 mg kg^{-1} in controls and 0.81 mg kg^{-1} in sludge-treated corn. Such crops are used only for animal feed (Peterson et al., 1979; Peterson et al., 1982). In a similar study (Hinesly et al., 1982) with high sludge loading rates, Zn, Cd, and Ni were increased in corn grain and leaves, but Cu, Cr, and Pb were not increased. Plant-soil concentration ratios indicated that metals were less readily available from the sludge than from the original spoil material. Even though the grain accumulated more Cd and Zn than controls, there was no increase in concentration of these elements over seven growing seasons as a result of repeated annual applications (Hinesly et al., 1979a). On sludge-amended spoils in Kentucky, Ni, Cr, Cu, Cd, Mn, and Fe concentrations were not significantly increased in corn and soybean plants at 112 or 269 Mg ha^{-1} sludge rates (Feuerbacher et al., 1980), nor did Cu and Zn in corn and bush beans increase as sludge rates increased on sand and gravel spoils (Hornick, 1982). Copper and Zn values were similar to controls, but Cd increased with the sludge rate. All values for corn grain were well below suggested tolerance levels (Cd 0.27 vs 3.0 mg kg^{-1}; Zn 32 vs 300 mg kg^{-1}; Cu 4.5 vs 150 mg kg^{-1} (Hornick, 1982), and corn was considered a better metal excluder than soybeans (Feuerbacher et al., 1980). In another study corn grown on spoils amended with 56 Mg ha^{-1} of sludge showed no increased accumulation of Hg or Pb in the grain, and the protein content was increased over the control corn. Zinc concentrations in the grain was higher for the control than for the sludge-treated corn. Most of the metals were concentrated in the germ fraction rather than in the endosperm, and the corn products produced were found to add only minor quantities of Cd, Pb, and Hg to the human diet (Blessin and Garcia, 1979).

Hinesly et al. (1984) investigated the effects of annual furrow irrigation of liquid digested sludge on a calcareous strip mine spoil in Illinois on Cd and Zn uptake by corn over seven years. Liquid sludge applications were 0, 6.4, 12.7, and 25.4 mm. Repeated applications resulted in additive increases of both Cd and Zn in corn leaves and grain.

Few attempts have been made to produce horticultural crops on spoils, but Tunison and others (1982) grew highbush blueberries on spoils amended with sludge containing 1250 mg kg^{-1} Zn. High Zn levels and low Mg foliar levels resulted in severe chlorosis, but when the sludge was composted with bark, the symptoms were alleviated. The berries showed no significant accumulation of Cd, Cr, Cu, Ni, Pb, or Zn, and these elements were within limits considered safe for human consumption (Chaney, 1973).

Table 59. **Elemental Concentrations in Plants in Relation to Suggested Tolerance Levels for Agronomic Crops[a]**

			Tolerance Level in Agronomic Crops		
Reference	Years	Species	Below	Above	Variable
Haghiri and Sutton, 1982	3	Tall fescue	Cd, Cu, Ni	Zn	
Seaker and Sopper, 1982	3	Tall fescue Orchardgrass Birdsfoot trefoil	Al, Zn, Cu, Co, Pb, Ni, Cd		Cr
Kerr et al., 1979	3	Reed canarygrass	Cu, Zn, Cd	Ni	
Fitzgerald, 1982	4	Forages	Cu, Ni, Zn	Cd, Cr	Pb
Hinesly et al., 1982	—[b]	Corn leaf	Ni, Zn, Cu, Cr, Pb		Cd
Griebel et al., 1979	_	Tall fescue Birdsfoot trefoil	Cu, Ni, Zn, Cd		
Stucky and Bauer, 1979	—	Tall fescue Switchgrass Orchardgrass	Cu, Ni, Pb	Zn, Cr	Cd
Mathias et al., 1979	—	Tall fescue	Cu, Zn, Cd, Ni		Pb,Cr
Hinkle, 1982	3	Korean lespedeza Tall fescue Weeping lovegrass	Cu, Cr, Cd, Ni, Pb, Zn		
Franks et al., 1982	2	Bermudagrass Switchgrass Kleingrass Bluestem	Cu	Zn, Cd	
Stucky and Newman, 1977	2	Alfalfa Tall fescue	Ni, Cd, Zn, Cu, Pb, Cr		
Hornick, 1982	—	Corn grain	Cd, Zn, Cu		
Tunison et al., 1982	—	Blueberries	Zn, Cu, Cr, Ni, Pb, Cd		
Feuerbacher et al., 1980	—	Corn plant Soybean plant	Ni, Cu, Cd, Mn	Cr	
Schneider et al., 1981	—	Hardwood leaves Pine needle	Cu, Zn, Ni		
Seaker and Sopper, 1984a	5	Tall fescue Birdsfoot trefoil	Zn, Pb, Cu Cd, Ni		
Seaker and Sopper, 1983	5	Tall fescue Birdsfoot trefoil Panic grass	Cu, Zn, Cr Ni, Co, Cd		Pb
Sopper and Seaker, 1990	12	Orchardgrass Crownvetch	Zn,Cu,Pb Ni,Cd		
Joost et al., 1987	4	Reed canarygrass Tall fescue Redtop	Ni	Cd, Cr Pb	
Roberts et al., 1988	3	Tall fescue	Zn, Cu, Cd		
Pietz et al., 1989b	3	Alfalfa Bromegrass Orchardgrass Tall fescue	Cu, Ni, Pb, Zn	Cd	
Sopper, 1990a	3	Crownvetch	Cu, Zn, Cd Ni, Pb		
Sopper, 1990a	3	Reed canarygrass	Cu, Zn, Cd Ni	Pb	
Sopper, 1991a	12	Crownvetch Birdsfoot trefoil Panic grass	Cu, Zn, Cd Ni, Pb		

Table 59.   Elemental Concentrations in Plants in Relation to Suggested Tolerance Levels for Agronomic Crops[a] (Continued)

Reference	Years	Species	Tolerance Level in Agronomic Crops		
			Below	Above	Variable
Sopper and Kerr, 1982	2	Tall fescue Orchardgrass Crownvetch Birdsfoot trefoil	Cu, Zn, Cd Ni, Pb		
Murray et al., 1981	2	Tall fescue Orchardgrass Crownvetch Birdsfoot trefoil Alfalfa	Cu, Zn, Cd Ni, Pb		
Sopper, 1991b	2	Tall fescue Orchardgrass Birdsfoot trefoil Crownvetch	Cu, Zn, Cd Ni, Pb		
Sabey et al., 1990	1	Sagebrush Fourwing saltbush	Cu, Zn, Pb Cu, Pb	Cd	Zn, Cd
Fresquez et al., 1990	2	Gramma Galleta Bottlebush squirreltail	Cu, Zn, Pb		
Dressler et al., 1986	2	Tall fescue Orchardgrass Birdsfoot trefoil	Cu, Zn, Cd Ni, Pb		
Alberici et al., 1989	2	Tall fescue Orchardgrass Birdsfoot trefoil Quackgrass	Cu, Zn, Cd Ni, Pb		
Coker et al., 1982	4	Fodder oats Perennial ryegrass	Cu, Zn, Cd Ni		
Metcalfe and Lavin, 1991	4	Grass/clover mixture (Nickerson red circle)	Cu, Zn, Cd Ni, Cr		
Metcalfe and Lavin, 1991	4	Grass mixture (Mommersteeg 22)	Cu, Zn, Cd Ni, Cr		
Hinesly and Redborg, 1984	3	Corn	Cu, Zn, Ni Cr, Pb, Mn	Cd	
U.S. Army Corps of Engineers, 1987	3	Kentucky bluegrass Tall fescue Red fescue Weeping lovegrass	Cu, Zn, Cd Pb, Ni, Mn		

[a] See Table 30 for tolerance levels.
[b] Length of study uncertain.

Table 59 lists some effects of sludge applications on the levels of trace metals in plant tissue found in various studies. Most studies found trace metal concentrations in the vegetation to be below tolerance levels for yield suppression and well below phytotoxic levels.

Summary

Sludge applications on mine land generally increase the concentrations of macronutrients (N, P, K, Ca, and Mg) in grass and legume species, field crops, and tree species. Concentrations of K in municipal sludge is usually low and thus the effect on plant concentrations is usually minimal. The greatest effect is usually a significant increase in foliar N concentrations.

Sludge applications generally result in an increase in plant concentrations of trace metals. These increases are usually the highest the first year after sludge application and then decrease with time. Most of the time these increased levels in trace metals in the plants are below the suggested tolerance level for agronomic crops and are not phytotoxic. Although trace metals are applied on the site with the sludge application, there is a dilution effect in terms of plant uptake because of the increased biomass. Concentration of trace metals in the vegetation are often below the required concentration for total dairy rations and are generally within the levels considered safe for animal consumption.

EFFECTS ON SOIL

Physical Properties

The high organic matter of sludge improves the physical condition of barren spoils tremendously. Parameters that benefit from sludge incorporation include water-holding capacity, bulk density, and surface temperatures. In Illinois, incorporation of digested sewage sludge at rates of 0, 224, 448, and 896 Mg ha^{-1} greatly improved the spoil physical properties (Hinesly et al., 1982). Ten months after sludge incorporation, untreated spoil contained 12.2% of water stable aggregates greater than 0.25 mm in diameter, as compared to 42.1% in spoil amended with 896 Mg ha^{-1} of sludge. The available water holding capacity was increased from 14.8% in untreated spoil to 21.1% with 896 Mg ha^{-1} of sludge.

In Colorado, Topper and Sabey (1986) reported that application of digested sewage sludge on mine spoils at rates of 0, 14, 28, 55, and 83 Mg ha^{-1} significantly increased saturation water percentages. Percent saturation was increased from 27.9% on untreated spoil to 35.6, 41.9, and 43.2% on the three highest sludge application rates, respectively. This indicates an increase in water-holding capacity due to the added organic matter from the sludge.

Composted sludge additions of 160 Mg ha^{-1} doubled the percentage of moisture in sand and gravel spoils (Hornick, 1982), and increased water retention of coal refuse (Joost et al., 1981) and mine spoils (Schneider et al., 1981). Increased volumes of percolate water were reported on the Palzo mine site in Illinois after liquid sludge was disked 15 cm into the spoils (Urie et al, 1982). On Ohio strip mine spoils (Haghiri and Sutton, 1982), however, the opposite effect occurred because of the rapid uptake and transpiration of soil water by increased plant biomass. In fact, the volume of percolate water was indirectly proportional to the sludge loading rate. Because sludge is about 50% organic matter (dry basis), it can improve water infiltration and retention (Griebel, et al., 1979). Younos and Smolen (1981), reporting on simulation of the infiltration process in sludge-amended mine spoil, found that sludge promotes infiltration, rapid saturation, and increased hydraulic conductivity.

Sludge additions have resulted in decreased bulk density in Illinois mine spoils high in compacted clay (Peterson et al., 1979; Peterson et al., 1982) and coal refuse (Joost et al., 1981); in decreased temperatures on sand and gravel spoils (Hornick, 1982); and in reduced runoff, erosion, and sedimentation from coal mine spoils (Sutton and Vimmerstedt, 1974).

Joost et al. (1987) reported that applications of sewage sludge at 225, 450, and 900 Mg ha^{-1} reduced bulk density of acid coal refuse (gob). Sewage sludge application increased organic matter content of the gob by 2.0 to 2.5 times that of unamended gob. The sludge applications also increased the number of water-stable aggregates and the proportion of macropores present in the gob. Sludge-treated gob exhibited significantly higher water content at 0.03 mPa than untreated gob.

Cocke and Brown (1987) investigated the effect of applications of 142 and 284 Mg ha^{-1} of sewage sludge on the physical properties and vegetation yields on lignite mine spoil in Texas. Water infiltration, aggregate stability, available soil water, and vegetation yields all increased significantly as a result of sludge application. Sewage sludge increased water infiltration by 38%, aggregate stability was 16% greater, and available soil water 18% higher than in control plots. Average bermudagrass yield increased significantly from 0.5 Mg ha^{-1} on the control plot to 4.0 Mg ha^{-1} on the 284 Mg ha^{-1} sludge-amended plot.

Summary

Sludge applications on mine land generally improves the physical properties of the spoil material making it a better growing medium for vegetation establishment and growth. Because sludge has a high organic matter content it increases the water holding capacity of the spoil, increases water infiltration capacity, decreases bulk density, increases saturation water percentages, and tends to reduce spoil surface temperatures. In addition it also tends to increase the number of water stable aggregates.

Chemical Properties

When sludges are applied to land, an increase in soluble salt content of the growing media may result. At a sludge application rate of 896 Mg ha^{-1}, as compared with a 224 Mg ha^{-1} rate, and an electrical conductivity (EC) of 6.6 mmho/cm, a 50% reduction in corn yield resulted. High soluble salts may also affect the establishment of some grass species (Hinesly, et al., 1982). Hinesly noted an increase in EC with increasing sludge levels (0, 224, 448, 896 Mg ha^{-1}) applied to a calcareous coal mine spoil in Illinois and speculated this caused a decline in vegetation growth. Electrical conductivities ranged from 2.2 mmho cm^{-1} in spoil samples from control plots to 6.6 mmho cm^{-1} in spoil amended with 896 Mg ha^{-1} of sludge. Topper and Sabey (1986) also reported significant increases in EC with applications of sludge at 28, 55, and 83 Mg ha^{-1} on coal mine spoil in Colorado. The greatest level of sludge application (83 Mg ha^{-1}) resulted in an observed mean EC of 5.5 dSm^{-1}. However, measurements of above ground biomass seemed to indicate there were no adverse salinity effects on plant growth.

On another site, however, EC was significantly reduced after sludge incorporation in very stony spoil that was high in Fe, Al, and Mn (Schneider et al., 1981).

Table 60.   Changes in Spoil pH Over a 12-Year Period on a Pennsylvania Strip
             Mine Spoil Bank Treated with 184 Mg ha⁻¹ of Sludge

Depth (cm)	Soil pH					
	May 1977[a]	Sept. 1977	Nov. 1978	Oct. 1979	May 1981	Aug. 1989
0–15	3.8	6.2	6.7	7.3	5.8	5.4
15–30	3.8	4.2	4.6	5.1	3.4	5.6

Source: Sopper and Seaker (1990).

[a] Presludge samples.

Lejcher and Kunkle (1974) also reported that after sludge incorporation into coal refuse EC at the 0 to 30 cm depth was decreased. Haynes and Klimstra (1975) reported that EC greater than 1.5 mmho cm⁻¹ may be damaging to crops. In another study, EC of sludge-amended spoils ranged from 0.68 to 2.80 mmho cm⁻¹, but vegetative cover was not affected as much by EC as it was by pH (Stucky and Bauer, 1979). Potential soluble salt problems can usually be avoided by proper management techniques. Maximum effective loading rates should be based on the constituents in the sludge and the crop that is to be grown (Hinesly et al., 1982).

Limed sludges have a neutral to alkaline pH and this can raise the pH of acid spoils (Griebel et al., 1979). The pH of coal refuse gob was increased from 2.6 to 5.3 even without limestone when sludge was incorporated at rates of 450 to 900 Mg ha⁻¹. Sludge was more effective than 180 Mg ha⁻¹ limestone in raising pH (Joost et al., 1981). Sludge applied at 67 to 269 Mg ha⁻¹ with lime increased pH considerably, but 34 Mg ha⁻¹ sludge with lime had no effect (Feuerbacher et al., 1980). Even after 195 Mg ha⁻¹ of lime and commercial fertilizer were applied to spoils in Ohio with a 2.3 pH, rye cover was considerably poorer than on adjacent plots amended with sludge only (Sutton and Vimmerstedt, 1974). When anthracite refuse was irrigated with liquid digested sludge and sewage effluent, significant increases in pH to a 76 cm depth occurred (Kardos et al., 1979). After six consecutive lime and sludge applications in four years, pyrite mine spoils increased in pH from 2.4 up to 5.0, and after the seventh application pH ranged from 6.1 to 6.6 (Hinkle, 1982).

Limestone initially increases the pH of acid spoils, but the pH eventually declines as sulfur-bearing minerals are oxidized (Sutton and Vimmerstedt, 1974). When commercial fertilizer plus lime was applied on Virginia acid mine spoils, pH remained below 4.0 for four years, but where municipal compost (pH 8.5) was added once without any lime, pH ranged from 4.9 to 7.4 through the four-year period (Scanlon et al., 1973). Little information is available on long-term effects of single high applications of sludge on mined land. However, Sopper and Seaker (1990) reported that a sludge application of 184 Mg ha⁻¹ increased the pH of strip mine spoil in Pennsylvania from 3.8 up to 6.2 in the surface 15 cm four months after application, and was still 5.4 after 12 years with excellent grass and legume cover (Table 60). (Sopper and Kerr, 1982, Sopper and Seaker, 1990). Sludge alone significantly raised the pH of spoils in West Virginia, and sludge plus lime was even more effective. Initial pH values of 3.0 to 4.0 were increased above 5.0 and usually above 6.0, with no decline after three years (Mathias et al., 1979). On Ohio strip mine spoil where two sludges were applied at 179, 358, and 716 Mg ha^{-1} the pH did not decline over a three-year period (Haghiri and Sutton, 1982). When Chicago sludge was applied annually to strip mines over a six-year period, the pH decreased slightly the first three years, mainly because of the nitrification and organic-acid production in the soil. The decrease was minimal, however, and the pH appeared to stabilize during the next three years (Peterson et al., 1982). Hinesly and others (1982), however, observed a drop in pH from 7.5 to 6.0 in calcareous spoil amended with 896 Mg ha^{-1} of Chicago sludge. Topper and Sabey (1986) reported similar declines in pH on Colorado mine spoil having an initial pH of 7.1. Applications of sludge at rates of 14, 28, 55, and 83 Mg ha^{-1} resulted in spoil pH's for the 0 to 15 cm depth of 6.8, 6.5, 6.3, and 6.2, respectively. It has been recommended that sludge amended spoil material be maintained at a pH of at least 6.5 so that potentially toxic metals will remain at relatively low solubility levels (Sopper and Seaker, 1983). Pietz et al. (1989a) also reported a pH decline with time after treating a coal refuse pile with 542 Mg ha^{-1} of sludge. The pH of the surface refuse (0 to 15 cm) was 2.8 on a control plot and was increased to 5.0 in 1976 following sludge applications. By 1981, the pH had dropped to 3.3.

In a zinc smelting area, lime and sludge raised soil pH, with an increase from 5.8 to 6.5 occurring within two years of the application (Franks et al., 1982). Soil pH is a critical factor for plant growth, and complete mixing of the lime and sludge is essential for uniform vegetative cover (Sutton and Vimmerstedt, 1974). Fibrous root systems, such as those produced by tall fescue, may have a stabilizing effect on pH, according to Stucky and Newman (1977).

Some trials have attempted to assess the effect of deep incorporation of sludge on plant establishment and growth, but results are vague. Incorporation of sludge to 60 cm as compared with a 30 cm incorporation, did not result in better vegetative cover the first year (Joost et al., 1981). In another study incorporation from 15 cm to a 40 cm depth increased pH at the deeper level, and deeper root penetration as well as better root quality resulted (Feuerbacher et al., 1980). Root systems are often confined to nontoxic or treated layers when sludge is incorporated into spoils (Sutton and Vimmerstedt, 1974), but are usually much more prolific and deeper than when inorganic fertilizer is used (Scanlon et al., 1973).

Cation exchange capacity (CEC) is normally improved by sludge addition to spoils because of the high CEC of organic matter (Stucky and Newman, 1977; Schneider et al., 1981; Jones and Cunningham, 1979). CEC is an important factor in the availability of cations, both nutrient (Ca, Mg) and trace metals (Cu, Zn, Pb, Cd), to plants and also affects their movement into groundwater.

The effect of sludge incorporation on the amount and availability of major plant nutrients varies. In general, sludges apply considerable N and P, but little K. The Ca, Mg, and S contribution varies with the composition of the sludge. Peterson et al. (1982) used soil N, P, and K as an index of soil rejuvenation by sludge, and found that over a four-year period repeated application of Chicago sludge increased available N, P, and K each year on both agricultural soils and

Table 61.   Chemical Analyses of the Refuse Bank Before and Three Years After
Sludge Application for pH, Kjeldahl Nitrogen, and Bray Phosphorus

Sludge Application (Mg ha⁻¹)	Refuse Depth	pH	Kjeldahl N (%)	Bray P (mg kg⁻¹)
Pre-treatment	cm			
0	0–30	5.1	0.140	193
75	0–30	4.7	0.016	266
0	30–60	5.6	0.048	320
75	30–60	5.1	0.008	398
0	60–90	5.2	0.064	198
75	60–90	5.4	0.082	231
Post treatment				
0	0–15	3.8	0.041	216
75	0–15	6.2	0.219	473
0	15–30	4.8	0.036	534
75	15–30	5.9	0.043	268
0	30–60	4.9	0.072	327
75	30–60	5.0	0.023	447

Source: Kerr et al. (1979).

mine spoil. Significant increases in Kjeldahl-N and Bray-P occurred in burned anthracite refuse where dewatered and heat-dried sludge was applied at 75 Mg ha⁻¹ in Pennsylvania. After three years, total N in control spoils for the 0 to 15 cm depth was 0.041% vs 0.219% in the sludge-treated spoils (Table 61). Phosphorus concentration was 216 mg kg⁻¹ in control spoils vs 473 mg kg⁻¹ in sludge-treated spoils (Kerr et al., 1979). On abandoned pyrite mines, commercial fertilizer and sludge applied together increased spoil phosphate and K_2O, but K availability was much greater where good vegetative cover existed than where the site was bare, indicating the importance of plants in the uptake and recycling of nutrients (Hinkle, 1982). Sludge additions increased spoil P, Ca, and Mg, but not K in West Virginia (Mathias et al., 1979). Spoil K decreased after three years, probably because of removal of vegetation. At the Palzo mine site in Illinois, only K remained deficient after sludge application (Jones and Cunningham, 1979). Total and extractable N and P were increased by four different sludge application rates in Kentucky spoils, but there was no change in K content (Schneider et al., 1981). Sutton and Vimmerstedt (1974), however, observed a threefold increase in available K and an eighteenfold increase in available P from applications of sludge at 658 Mg ha⁻¹. In another study extractable P was greater in sludge-amended spoils than in controls, but an increase in sludge loading from 112 to 224 Mg ha⁻¹ did not further increase P. It is possible that the increased Ca at the higher rate neutralized some of the acid in the P extractant, since both Ca and Mg were markedly increased by sludge additions (Mathias et al., 1979).

In New Mexico, Fresquez et al. (1990a, 1990b) reported on a study where dried, anaerobically digested sludge was applied at 0, 22.5, 45, and 90 Mg ha⁻¹ to a degraded semiarid grassland site to determine the effects on soil chemical properties. They reported that most soil macronutrients, such as N, P, and K, and micronutrients, such as Cu, Fe, Mn, and Zn increased linearly with increasing sludge amendment rates. Trace metals, Cd and Pb, did not change as a result of sludge amendment in the first three growing seasons. However, concentrations of soil Cu, Mn, and Cd were just above maximum acceptable levels in the heaviest sludge treatment after four growing seasons. Soil pH decreased from 7.8 to 7.4 with the application of sludge after the second growing season, but did not significantly increase the solubility of soil trace metals.

Interpretation of reports on metal loadings to disturbed land after sludge application is difficult, mainly because the analytical methods used to determine metal concentrations are inconsistent. Values for extractable vs total metals are quite different, as are those for HCl-extractable vs DTPA-extractable metals. Roth and others (1982) found that Cd, Cu, Mn, and Ni concentrations were significantly different when four extractants were compared: DTPA (pH 7.3; pH 4.9) and 0.1 N HCl (pH 1.3; pH 4.9). No extraction method has yet been developed that adequately indicates the amount of metals available to plants. Plant uptake may be much less than indicated by DTPA extraction (Griebel et al., 1979). Such factors as pH, organic matter content, metal concentration of sludge, phosphorus, and iron concentrations, and CEC all contribute to plant availability of trace metals, and further complicate the matter. For any one study, however, a comparison can be made of spoil metal concentrations before and after sludge additions, on the basis of a single known method of analysis.

Sludge addition usually results in increases in spoil heavy metal content, but which metals increase depends on the particular sludge, and in many cases, the increases are not significant. Sludge organic matter and pH effects often cause a decrease in availability of Fe, Al, and Mn, which are already found at extremely high concentrations in most acidic spoil materials. On anthracite refuse, liquid-sludge irrigation greatly increased phosphorus, which tended to tie up and detoxify Fe, Al, and Mn in the upper root zone (Kardos et al., 1979). In Pennsylvania dewatered sludge incorporation of 184 Mg ha[-1] reduced extractable Fe, Al, and Mn in the strip mine spoil plow depth but increased Cu, Zn, Cr, Pb, Cd, and Ni. The increases were minimal, however, and sludge did not affect concentrations below a 15 cm depth (Sopper and Kerr, 1982; Seaker and Sopper, 1983, 1984).

Application of dewatered heat-dried sludge at 75 Mg ha[-1] on a burned anthracite refuse bank had minimal effect on trace metal concentrations in the refuse. Concentrations of Cu, Zn, Pb, Cd, and Ni were all increased in the 0 to 15 cm depth; however, the concentrations were all in the normal range for Pennsylvania soils, with the exception of Cd, which was slightly higher (Table 62).

More recently, Sopper (1991b) reported on the effects of an application of a sludge-fly ash mixture on a steep coal mine refuse bank on the chemical properties of the refuse two years after treatment.

Analyses of refuse samples collected prior to treatment showed that, in general, the chemical characteristics of the refuse were quite similar on the three subplots of each treatment. Analyses of post-treatment samples collected two years after treatment show little change in the chemical status of the refuse in the 0 to 60 cm depth (Table 63). There were no significant differences in pH, TKN, P, K, Mg, Ca, or total soluble salts in the refuse of both the control and amended plots at all three

Table 62.   Chemical Analyses of the Refuse Bank Before and 3 Years after Sludge Application for Extractable Trace Metals

Sludge Application (Mg ha⁻¹)	Refuse Depth (cm)	mg kg⁻¹				
		Cu	Zn	Pb	Cd	Ni
Pre-treatment						
0	0–30	7.0	3.5	0.33	0.02	1.21
75	0–30	6.5	2.0	0.80	0.02	0.53
0	30–60	6.5	3.5	0.55	0.02	0.44
75	30–60	7.0	3.0	0.98	0.03	0.63
Post-treatment						
0	0–15	3.5	2.0	4.48	0.07	0.93
75	0–15	72.0	118.0	71.41	0.85	3.79
0	15–30	4.0	1.5	1.38	0.02	0.74
75	15–30	2.0	3.5	0.63	0.02	0.81
0	30–60	2.0	2.0	0.79	0.01	1.00
75	30–60	2.5	6.5	1.83	0.06	0.79
Normal Range in Pennsylvania Soils[a]	2–	10–100	2–300	0.01–200	5–0.70	500

[a]   From Baker and Chesnin (1975); Kerr et al. (1979).

depths. Phosphorus concentrations increased dramatically on both sets of plots over pretreatment values. On the control plots P concentration increased from 11.3 to 63.7 mg kg⁻¹ and on the sludge-fly ash amended plots P increased from 6.0 to 68.5 mg kg⁻¹. Average concentrations of Mn, Fe, and Al in the refuse were quite similar on both sets of plots at all three depths (Table 64). Only Fe and Al concentrations were significantly higher at the 30 to 60 cm depth on the amended plots. Concentrations of trace metals increased slightly on the sludge-fly ash amended plots but only concentrations of Zn at the 0 to 15 cm depth was significantly higher (Table 65). It appears that precipitation has not yet leached many of the constituents from the sludge-fly ash amendment into the refuse.

In Fulton County, Illinois, Peterson and others (1982) reported that Chicago sludge, which is very high in metals, increased spoil Cd, Zn, Ni, and Cu. When applied annually, however, only Cd exceeded the values for normal ranges in

Table 63.   Chemical Analyses of the Refuse 2 Years After Treatment

Treatment	Refuse depth (cm)	pH	TKN (%)	P (mg kg⁻¹)	K (meq 100g⁻¹)	Mg (meq 100g⁻¹)	Ca (meq 100g⁻¹)	T.S.S. (mmho cm⁻¹)
Control	0–15	5.9	0.45	63.7	0.11	0.9	1.2	0.10
(Fertilizer)	15–30	5.5	0.45	34.5	0.09	0.8	1.0	0.10
	30–60	5.1	0.44	16.0	0.08	0.6	0.8	0.10
Amended	0–15	5.6	0.43	68.5	0.04	0.7	1.8	0.17
(2S:1FA)	15–30	5.0	0.47	15.2	0.04	0.6	1.2	0.13
	30–60	4.6	0.45	7.8	0.03	0.5	0.9	0.10

Source: Sopper (1991b).

Table 64. **Average Concentrations of Available Mn, Fe, and Al in the Refuse 2 Years After Treatment**

Treatment	Refuse Depth (cm)	mg kg⁻¹		
		Mn	Fe	Al
Control	0–15	1.1	22	15
(Fertilizer)	15–30	1.0	22	18
	30–60	1.7	25	24
Amended	0–15	1.4	32	22
(2S:1FA)	15–30	0.9	37	32
	30–60	2.4	35[a]	38[a]

Source: Sopper (1991b).

[a] Significant effect at $P < 0.05$.

agricultural soils reported by Allaway (1968). Another study (Fitzgerald, 1982) with the same sludge found large increases in Cd, Zn, Ni, Cu, Cr, Pb, and Hg in spoils as compared with controls. Copper, Zn, and Cd were at times above reported normal ranges. However, except for Zn and Pb, the concentrations did not appreciably increase over the four years of annual application. Unamended gob material may be high in Mn and Cr (Joost et al., 1981), and sludge additions to gob were found to cause large increases in Cr, Mn, Cu, Zn, Cd, and Pb based on maximum soil concentrations reported by Yopp et al. (1974) that might exert toxic effects on plants. However, grass establishment and growth apparently were not affected. Extractable Cu, Ni, Zn, and Cd increased as compost application rates increased on Maryland mine spoils (Griebel et al., 1979). Phosphate rock, which can contain significant trace metals, raised the values even higher, but limestone additions decreased them. Zinc, Cu, Pb, and Ni may increase in availability and potential toxicity as pH decreases (Council for Agricultural Science and Technology, 1976).

In some instances, the availability of metals in spoil was not increased by sludge additions, due to the metal composition of the sludge and increases in pH

Table 65. **Average Concentrations of Available Trace Metals in the Refuse 2 Years After Treatment**

Treatment	Refuse Depth (cm)	Trace Metals (mg kg⁻¹)					
		Cu	Zn	Pb	Ni	Cd	Cr[a]
Control	0–15	1.3	0.4	0.3	0.6	0.02	5.9
(Fertilizer)	15–30	1.6	0.5	0.3	0.5	0.01	5.8
	30–60	2.1	0.6	0.4	0.3	0.08	6.1
Amended	0–15	3.4	2.3[b]	0.5	0.3	0.04	8.4
(2S:1FA)	15–30	1.5	0.6	0.4	0.3	0.01	5.6
	30–60	1.4	0.6	0.7	0.4	0.02	5.6

Source: Sopper (1991b).

[a] Chromium values are for total metal concentrations.
[b] Significant effect at $P < 0.05$.

Table 66. **Changes in Concentrations of Kjeldahl-Nitrogen, Bray-Phosphorus, and Exchangeable Cations in the Spoil Collected at the 0–15 cm Depth on a Sludge-Amended Mine Site in Pennsylvania**

Year	Kjeldahl Nitrogen (%)	mg kg⁻¹			
		Bray Phosphorus	K	Ca	Mg
May 1977[a]	0.04	2	12	541	452
Sept 1977	0.05	11	19	1222	32
1978	0.09	9	23	2600	40
1979	0.16	38	46	3873	53
1981	0.34	79	45	1298	99
1984	—	91	74	1440	108
1989	0.12	83	30	733	84

Source: Sopper and Seaker (1990).

[a] Presludge samples.

and organic matter. For example, in soils contaminated by zinc smelting, Zn, Cu, and Cd levels were extremely high and sludge addition improved plant cover without significantly adding to the metal load (Franks et al., 1982). On abandoned pyrite mines a dramatic drop in metal concentrations occurred as reclamation progressed, probably because of increases in spoil pH (Hinkle, 1982). Concentrations of Cu, Fe, Mn, and Zn were within normal soil ranges reported by Allaway (1968). Hinesly and others (1982) observed no increase in As, Mo, or Mn in sludge-amended spoils, due to low concentrations of these elements in the sludge.

Not much information is available relative to the long-term effects of single applications of sewage sludge on mine spoil or coal refuse. However, recent publications by Sopper and Seaker (1990) and Sopper (1991) provides some insight into long-term effects on spoil and refuse chemical properties. In the study, dewatered sludge was applied at 184 Mg ha⁻¹ on an abandoned strip mine spoil bank in 1977. Soil samples were collected over a five-year period (1977 to 1981) and then again in 1984 and 1989. Changes in concentrations of Kjeldahl-nitrogen, Bray-phosphorus, potassium, calcium, and magnesium are given in Table 66. The nutrient status of the spoil shows a general increase in concentrations of Kjeldahl-N up to 1981 and up to 1984 for Bray-phosphorus, K, and Ca. The application of lime and sludge initially resulted in a decrease in the concentration of Mg; however, since 1978 there has been a steady increase. The 1989 values are lower but still quite adequate to support plant growth.

Concentrations of extractable trace metals in the 0 to 15 cm spoil depth are given in Table 67. Concentrations of Cu, Zn, Cr, Pb, Cd, and Ni all show a steady increase for the first five years (1977 to 1981). By this time, all the sludge organic matter was probably mineralized and all trace metals released to the surface spoil. Results of spoil analyses in 1984 and 1989 show a gradual decrease in concentrations of all trace metals.

Although the sludge application did increase the concentrations of extractable trace metals in the 0 to 15 cm spoil depth, these higher concentrations are still within the normal ranges for these elements in U.S. soils (Allaway, 1968).

Table 67.   Changes in Concentrations of Extractable Trace Metals from Spoil
            Collected at the 0–15 cm Depth on a Sludge-Amended Mine Site in
            Pennsylvania

Sampling Date	Trace Metals (mg kg⁻¹)					
	Cu	Zn	Cr[a]	Pb	Cd	Ni
May 1977[b]	2.5	2.9	0.2	0.5	0.02	1.1
Sept 1977	10.8	7.7	0.4	3.5	0.04	0.9
1978	8.8	7.7	0.2	2.3	0.02	1.2
1979	58.7	56.9	1.7	13.0	0.27	1.5
1981	87.3	74.6	3.5	22.7	0.95	2.8
1984	57.6	59.6	—	14.8	0.56	2.8
1989	51.9	37.8	—	13.5	0.42	2.0
Normal Range for U.S. Soils[c]	2–100	10–300	5–3000	2–200	0.01–7.00	5–500

Source: Sopper and Seaker (1990).

[a] Values for Cr are total concentrations.
[b] May 1977 values represent pretreatment conditions.
[c] Allaway (1968).

Analyses of spoil samples collected from the 15 to 30 cm depth showed a general increasing trend in trace metal concentrations from 1977 to 1989, indicating that some leaching of the trace metals through the spoil profile was occurring (Table 68).

In another long-term study, Seaker and Sopper (1983) and Sopper (1991) applied 80 and 108 Mg ha⁻¹ of dewatered sludge along with 11 Mg ha⁻¹ agricultural lime, to an anthracite coal refuse bank in northeastern Pennsylvania.

Samples of vegetation, soils, and groundwater were collected annually over a five-year period (1978 to 1982). In August, 1990 the site was revisited and refuse samples were collected at the 0 to 15 cm depth to evaluate long-term effects.

Addition of lime and sludge raised the initial pH of the mine refuse material (3.6 to 3.8), so that four years after incorporation the pH was 5.9 and 5.8 in the 0 to 15 cm depth on the 80 and 108 Mg ha⁻¹ plots, respectively (Table 69). The bulk of the vegetation root systems were located within this depth. Three years after incorporation, the spoil pH ranged from 6.8 to 7.8 in the top 5 cm. By 1990, the pH decreased slightly to 5.6 and 5.4 in the 0 to 15 cm depth in the sludge

Table 68.   Changes in Concentrations of Extractable Trace Metals from Spoil
            Collected at the 15–30 cm Depth Following Sludge Application

Sampling Date	Trace Metals (mg kg⁻¹)					
	Cu	Zn	Cr[a]	Pb	Cd	Ni
May 1977[b]	3.0	2.4	0.10	0.6	0.020	1.0
Sept 1977	4.0	2.0	0.10	1.3	0.010	0.4
1978	2.5	1.7	<0.01	1.3	0.007	0.7
1979	9.2	8.7	0.28	2.4	0.026	0.2
1981	2.4	2.8	0.05	0.5	0.014	0.4
1989	13.8	10.2	0.43	3.8	0.122	1.9

[a] Values for Cr are total concentrations.
[b] May 1977 values represent pretreatment conditions.

Table 69.   Mean pH of Anthracite Refuse Material
Collected at the Surface Before and After
Application of Sludge

| Year | Depth (cm) | Sludge Applied (Mg ha⁻¹) | |
		80	108
1978[a]	0-15	3.8	3.6
1979	0-15	3.8	3.6
1981	0-5	6.8	7.8
1982	0-15	5.9	5.8
1990	0-15	5.6	5.4

Source: Sopper (1991).

[a] Pretreatment values.

amended plots, respectively. The decrease in pH is probably the reason most of the birdsfoot trefoil plants have disappeared. Birdsfoot trefoil does not grow well if the pH is below 5.5.

Results of the mine refuse analyses for Kjeldahl-N, Bray-P, K, Ca, and Mg before and after sludge application of 108 Mg ha⁻¹ are given in Table 70. There appeared to be little change in N after sludge incorporation during the first five years. However, by 1990 the N concentration in the refuse was higher by almost 50%. The amount of Bray phosphorus was increased considerably by the sludge addition during the first five years and was more than doubled by 1990. Calcium and Mg concentrations more than double during the first five years, but sludge did not effect the K content of the refuse. Results of the 1990 refuse samples analyses indicated that phosphate (P_2O_5) and magnesium (MgO) levels in the plow layer are in the optimum range for plant growth. Average phosphate amount was 193 kg ha⁻¹ in comparison to an optimum range of 157 to 258 kg ha⁻¹. Average magnesium amount was 736 kg ha⁻¹ in comparison to an optimum range of 246 to 739 kg ha⁻¹. Potash was the only nutrient in the low range with 84 kg ha⁻¹ in comparison to an optimum range of 210 to 340 kg ha⁻¹.

Concentrations of extractable trace metals at the 0 to 15 cm depth showed little effect of sludge one year after the application, but by the fourth year (1982)

Table 70.   Changes in Concentrations of Kjeldahl Nitrogen, Bray Phosphorus,
and Exchangeable Cations from Anthracite Refuse Collected at the 0–
15 cm Depth Following Sludge Application at 108 Mg ha⁻¹

| Sampling Date | Kjeldahl Nitrogen (%) | mg kg⁻¹ | | | |
		Bray Phosphorus	K	Ca	Mg
1978[a]	0.41	7	27	470	45
1979	0.47	8	24	620	49
1982	0.46	16	19	1100	104
1990	0.65	37	42	1060	204

Source: Sopper (1991).

[a] Pretreatment samples.

Table 71.   Changes in Concentrations of Extractable Trace Metals from Anthracite Refuse Collected at the 0–15 cm Depth Following Sludge Application at 108 Mg ha[-1]

Sampling Date	Concentrations (mg kg[-1])						
	Cu	Zn	Cr	Pb	Co	Cd	Ni
1978[a]	7	1	<0.01	1.1	0.38	0.500	0.01
1979	7	3	0.05	0.9	0.35	0.003	0.97
1982	18	25	0.59	5.6	0.72	0.372	1.37
1990	30	59	1.91	31.7	0.83	0.970	2.20
Mean concentrations for U.S. soils[b]	30	57	—	17	—	0.270	24

Source: Sopper (1991).

[a]   Pretreatment samples.
[b]   Sommers et al. (1987).

increases ranged from 2-fold for Cu and Co to 100-fold for Ni (Table 71). However, the actual concentrations of these metals were very low, and the Cd concentration was decreased following sludge application. By 1990, all trace metal concentrations in the refuse had increased substantially. However, these increases were still minimal, and concentrations of Cu, Zn, and Ni were still below or about equal to average values for native soils in the United States (Sommers et al., 1987). Only Pb and Cd concentrations were considerably higher than mean concentration values reported for native soils in the United States.

Pietz et al. (1989a) reported similar results when sludge and lime were applied on coal refuse material in Illinois. The sludge and lime applications were 542 Mg ha[-1] and 89.6 Mg ha[-1], respectively. Results of coal refuse samples (0 to 15 cm) analyzed over a five-year period are given in Table 72. The lower water-soluble Al and Fe concentrations appeared to reflect the ability of the applied sewage sludge to retain Al and Fe, and the ability of lime to decrease the solubility of these metals through an increase in pH and precipitation of Al and Fe compounds. Sommers et al. (1984) and Corey et al. (1987) reported that the primary mecha-

Table 72.   Concentrations of Water Soluble Al and Fe in Coal Refuse Samples Collected from 1976 to 1981

	Treatment	Al (mg kg-1)	Fe (mg kg-1)
1976	Control	279	233
	Sludge and Lime	9	8
1978	Control	255	190
	Sludge and Lime	3	1
1979	Control	136	119
	Sludge and Lime	39	4
1980	Control	103	68
	Sludge and Lime	2	1
1981	Control	105	31
	Sludge and Lime	2	<1

Source: Pietz et al. (1989a).

nisms involved in the retention of metals by sludges are precipitation as carbonates, phosphates, sulfides, silicates, or hydrous oxides, and sorption by organic matter and hydros oxides.

Another long-term study has been reported by Palazzo and Reynolds (1991). They evaluated metal distributions in the upper 60 cm of acidic soil in a dredge disposal area which received 100 Mg ha^{-1} of anaerobically digested and dewatered primary sludge and 23 Mg ha^{-1} of lime. Soil samplings were collected at 2, 4, and 16 years after sludge application. The site was located along the Chesapeake and Delaware Canal in Delaware. The sludge had extremely high concentrations of metals (mg kg^{-1}): Cr (9584), Cu (2772), Zn (3470), Pb (1327), Ni (227), and Cd (27.3). Soil pH increased from 2.4 to 4.0, 4.8, and 7.4 after 2, 4, and 16 years, respectively. They reported that total soil metals concentrations in the 0 to 20 cm depth increased dramatically two years after sludge application (Table 73), but these levels were still near the ranges normally found in native soils (Allaway, 1968). Total metal concentrations were much lower in subsequent samplings at 4 and 16 years after sludge application. Between the second and fourth year all soils metals, except Cd, declined an average of 49 percent, with a range of 26 to 65%. By 16 years after sludge application, total soil metal concentrations had consistently declined by more than 86%. The concentrations of extractable metals in the 0 to 20 cm depth declined to very low levels 16 years after sludge application. Metals movement to lower soil depths appeared to be minimal.

Summary

Sludge applications on mine land generally has a beneficial effect on the chemical properties of the spoil material. The combined effect of applications of lime and sludge usually raises the pH of the spoil making it a better growing medium for vegetation. The high organic matter content of sludge usually increases the CEC of the spoil material which aids in the immobilization of trace metals. Sludge applications usually result in increases in trace metal concentrations in the spoil material, however, these increases are usually not significant. Trace metal concentrations usually reach a maximum after all the sludge is mineralized and then tend to decrease with time. Most research results show that trace metal concentrations in the spoil material following sludge applications are within the range considered normal for total and extractable metals in unpolluted and unamended soils. Sludge applications also generally increase the concentrations of macronutrients (N, P, Ca, and Mg) in the spoil material.

Biological Properties

Although the immediate goal of reclamation is to establish a vegetative cover that will prevent soil erosion, the long-term goal is soil ecosystem development and stability. Minespoils lack microbial activity and organic matter (Visser, 1985; Mills, 1985; Fresquez and Lindemann, 1982). Microbial processes such as humification, soil aggregation, and N cycling are essential in establishing productivity in minespoils, and productivity should be evaluated not only on aboveground

Table 73.   Total and Extractable Trace Metals in the Soil at Various Depths at Various Years After Sludge Application

	Copper			Zinc			Chromium		
Depth (cm)	2	4	16	2	4	16	2	4	16
Total									
0–20	201	95	11	306	227	44	380	133	46
20–40	26	14	17	57	46	49	51	50	35
40–60	21	9	17	71	55	12	13	35	18
Extractable									
0–20	66	41	<1	57	66	<1	29	<1	<1
20–40	12	4	<1	29	9	<1	6	<1	<1
40–60	6	2	<1	23	7	<1	3	<1	<1

	Nickel			Lead			Cadmium		
Depth (cm)	2	4	6	2	4	6	2	4	6
Total									
0–20	57	36	5	129	61	6	2.2	3.2	<2.0
20–40	23	36	4	32	20	4	0.3	2.8	<2.0
40–60	26	31	5	25	23	2	0.3	2.8	<2.0
Extractable									
0–20	30	5	<1	0.1	26	1.0	1.1	0.8	<2.0
20–40	13	11	<1	1.6	1.1	1.0	0.3	0.1	<2.0
40–60	15	9	<1	0.1	0.9	1.2	0.1	0.1	<2.0

Source: Palazzo and Reynolds (1991).

Note: All measurements are given in mg kg^{-1} for intervals of 2, 4, and 16 years for copper, zinc, and chromium; 2, 4, and 6, years for nickel, lead, and cadmium.

biomass, but also on the degree of development of functional microbial populations resembling those of an undisturbed soil. Microbial processes are so important to ecosystem recovery that the activity of microorganisms may be used as an index of the progress of soil genesis in minespoils (Schafer et al., 1980; Segal and Mancinelli, 1987).

If the premining organic layer (O horizon) has been destroyed, the only C source for microbial utilization is the plant biomass that is expected to accumulate over several growing seasons on the site. Until such accumulation occurs, microbial activity remains at a low level, with little improvement of adverse soil physical and nutrient conditions. Vegetation growth and maintenance are also inhibited. On sites reclaimed with chemical fertilizers and lime, vegetation may initially be established, but poor physical conditions result in deterioration of the vegetative cover before it can begin to ameliorate the spoil (Stroo and Jencks, 1982). On both alkaline and acidic minespoils, microbial activity, nutrient cycling, and spoil organic matter levels may take over 30 years to be reestablished (Segal and Mancinelli, 1987; Stroo and Jencks, 1982; Mills, 1985; Anderson, 1977; Schafer et al., 1980).

The use of sewage sludge as an organic amendment for minespoil reclamation has been extremely successful (Varanka et al., 1976; Frequez and Lindemann, 1982; Visser, 1985; Seaker and Sopper, 1984) because of its immediate improvement of spoil chemical and physical conditions, acceleration of plant establishment and growth, and achievement of long-term productivity. The organic C and nutrient content of sludge is responsible for achieving a self-maintaining cover on minespoils, but very few studies have quantitatively measured the effects of sludge application on microbial populations and activity, compared to sites reclaimed with lime and chemical fertilizer.

It has been hypothesized that heavy metals, many of which may be present in sludge, could potentially disturb the population dynamics and general ecology of soil microbes in natural habitats (Babich and Stotsky, 1977a). At high levels, inorganic salts of Zn, Cu, Cd, Cr, and Pb have been shown to interfere with microbial metabolism in laboratory cultures. Most studies involved metal concentrations far in excess of those found in land application systems using "typical sludges" with median metal concentrations at agricultural rates (Mathur et al., 1979; Bhuiya and Cornfield, 1974; Lighthart et al., 1983). Numerous studies have indicated that binding of metals to organic materials and clay minerals, precipitation, complexation, and ionic interactions significantly decrease their inhibitory effects on microbial activity (Gadd and Griffiths, 1978), so that inhibition by metals is substantially less in a soil system than in pure culture media (Babich and Stotsky, 1977b; Tomlinson, 1966).

A study on the Palzo tract in southern Illinois (Sundberg et al., 1979) found fungal populations in unreclaimed spoil to be only 1 to 2% of those in unmined agricultural soils. Application of sludge, and particularly incorporation, resulted in a tenfold increase in fungal activity, because of increase in pH and food supply, and better soil-moisture retention. Some fungi are introduced with the sludge, but with improved chemical and physical condition of the spoil and a vegetative cover that recycles organic matter and nutrients, natural successional changes and eventual stabilization of the fungal populations should occur.

A recent study reported by Seaker and Sopper (1988a) sheds some light on the value of sludge applications on mined land on the rejuvenation of microbial populations and activity. They conducted a field study of five strip mine sites reclaimed with sewage sludge and one site reclaimed by conventional methods (chemical fertilizer and lime) to assess the effects of sludge amendments and time on populations of bacteria, fungi, and actinomycetes, and on microbial respiration and organic matter decomposition. The sludge-amended sites ranged in age from one to five years following sludge applications at rates of 120 to 134 Mg ha[-1] (Table 74). The sludge amendment was from Philadelphia and consisted of a

Table 74.   Site Descriptions for Pennsylvania Microbial Community Study

Site	Age (yr)	Amendment	Application Rate[a] (Mg ha^{-1})	Date of Application	Lime Application (Mg ha^{-1})	pH (1985)[b]
1	1	Sludge	120	Sept. 1984	18	6.9
2	2	Sludge	128	June 1983	18	7.0
3	3	Sludge	128	May 1982	12	6.8
4	4	Sludge	134	July 1981	18	6.7
5	5	Sludge	134	July 1980	11	7.3
Fertilizer-amended	5	Fertilizer (23-10-20, N-P-K)	0.5	Aug. 1980	11	6.3

Source: Seaker and Sopper (1988a).

[a]   Dry weight basis.
[b]   At time of study.

mixture of composted sludge (with wood chips) and digested dewatered sludge cake. The mean and range of the concentration of constituents in sludge samples collected at the time of application on the five sites is given in Table 75 and the amounts of trace metals applied is given in Table 76.

Aerobic Heterotrophic Bacteria

Seaker and Sopper (1988a) reported that bacterial populations on the sludge-amended sites ranged from 4 to 63 × 10^6 g^{-1} (Table 77). Bacterial counts were 5 to 15 times higher on Site 1 than on the older sites, and were dramatically increased on all sludge-amended sites compared to the fertilizer-amended site. The first-year peak and subsequent stabilization of bacterial populations is a typical response following organic matter additions to minespoils (Fresquez and Aldon, 1986). Considering the extremely low initial pH of the minespoils in this study, commonly ranging from 3.0 to 5.0 prior to lime additions, the microbial populations achieved with lime and sludge amendments after only one year are remarkably high. They compare favorably with estimates of 1 to 34 × 10^6 g^{-1} reported for undisturbed soils (Wilson, 1965; Segal and Mancinelli, 1987; Miller and Cameron, 1978; Alexander, 1977; Visser, 1985; Miller and May, 1979).

Fungi

Sludge application resulted in fungal populations in the range of 4 to 18 × 10^5 g^{-1} (Table 77) (Seaker and Sopper, 1988a). These compare favorably with fungal populations in undisturbed soils which have been reported to range from 0.05 to 9 × 10^5 g^{-1} (Miller and Cameron, 1978; Segal and Mancinelli, 1987; Alexander, 1977; Wilson, 1965; Miller and May, 1979). Fungal numbers were three to four times higher on Site 1 than on the older sites, and were greatly increased on all sludge-amended sites compared to the fertilizer-amended site. Two other studies

Table 75.   Mean and Range of the Concentrations of Constituents in Sludge Samples Collected at Time of Application on Five Sites[a]

Constituent	Mean Concentration (mg kg⁻¹)	Range (mg kg⁻¹)
NO_3–N	260	89–493
NH_4–N	645	193–1,316
Organic–N	4,670	3,667–6,575
Total N	5,755	4,456–7,980
Total P	11,531	8,408–14,366
K	1,032	789–1,319
Ca	14,415	13,049–15,920
Mg	8,277	6,961–10,041
Na	460	367–613
Fe	19,272	18,010–20,017
Al	15,946	13,085–18,058
Mn	1,036	920–1,093
Zn	1,921	1,494–2,236
Cu	753	600–809
Cr	506	395–580
Pb	542	454–588
Ni	157	106–198
Co	24	7–38
Cd	6	4–8
Hg	0.5	0.5–0.6
Solids, %	48	47–49
pH	7.3	6.6–8.0

Source: Seaker and Sopper (1988).

[a]   Dry weight basis.

Table 76.   Amounts of Trace Metals Applied on Sludge-Amended Sites

Site	Trace Metals (kg ha⁻¹)					
	Zn	Cu	Cd	Pb	Ni	Cr
1	171	68	1.2	36	8	40
2	162	69	1.6	36	11	47
3	112	53	1.3	26	10	49
4	170	62	0.6	45	12	41
5	197	79	0.6	60	13	53
Mean	162	66	1.1	41	11	46
PDER[a] maximum	224	112	3.0	112	22	112

Source: Seaker and Sopper (1988a).

[a]   Pennsylvania Dept. of Environmental Resources.

Table 77.   Microbial Populations on Strip Mine Sites 1 to 5 Years Following
Sludge Application, and on Fertilizer-Amended Site

Site	Aerobic Heterotrophic Bacteria (10^6 g^{-1})	Fungi (10^5 g^{-1})	Actinomycetes (10^4 g^{-1})
1	63.67 ± 16.93a[a]	18.14 ± 5.45a	1.48 ± 1.04b
2	7.07 ± 1.32b	5.80 ± 2.35b	9.75 ± 5.48b
3	4.09 ± 0.77b	5.54 ± 1.32b	56.21 ± 26.71ab
4	11.37 ± 3.64b	3.98 ± 0.46b	140.23 ± 59.57a
5	13.74 ± 3.58b	4.03 ± 1.05b	40.89 ± 22.68b
F value	—[b]	—[c]	—[d]
Fertilizer-amended	3.06 ± 1.17	0.16 ± 0.04	6.94 ± 4.01

Source: Seaker and Sopper (1988a).

Note: Data reflects mean of 6 samples with standard error.

[a]   Means followed by different letters are significantly different at the 0.05 level of probability by the Waller-Duncan k-ratio t-test.
[b,c,d]   Significant effect at $P < 0.001$, 0.01, and 0.05, respectively.

on reclamation with sludge failed to find increases in fungal numbers, but did report increased species diversity (Fresquez and Lindemann, 1982; Parkinson et al., 1980).

Actinomycetes

Sludge applications resulted in actinomycete populations ranging from 1.48 to 140.23 × 10^4 g^{-1} (Table 77) (Seaker and Sopper, 1988a), compared to actinomycete populations for undisturbed soils, reported in the range of 1 to 436 × 10^4 g^{-1} (Alexander, 1977; Miller and Cameron, 1978; Visser, 1985; Miller and May, 1979; Segal and Mancinelli, 1987; Wilson, 1965). Actinomycetes exhibited a different pattern of development than the bacteria and fungi. These microbes are less competitive than the other groups and their populations were significantly lower on Sites 1 and 2 than on the older sites. The pattern follows that described by Alexander (1977), whereby the bacteria and fungi proliferate initially upon the addition of organic matter to the soil, and the actinomycete responses do not occur until later stages of decay, when competition has decreased. Actinomycete populations on Sites 3, 4, and 5 were considerably higher than on the fertilizer-amended site.

Nitrifying Bacteria

Seaker and Sopper (1988a) found that *Nitrosomonas* populations were not significantly different on the five sludge-amended sites (Table 78), but were two to four orders of magnitude greater than on the fertilizer-amended site. *Nitrobacter* had a significantly larger population on Site 1 than on the older sludge-amended sites, and were four to six orders of magnitude greater than on the control site. Populations of both genera ranged from 0.53 × 10^4 g^{-1} to 126.53 × 10^5 g^{-1},

Table 78.   Populations of Nitrifying Bacteria on Strip Mine Sites 1 to 5 Years Following Sludge Application, and on Fertilizer-Amended Site

Site	Nitrosomonas (10⁴ g⁻¹)	Nitrobacter (10⁵ g⁻¹)
1	29.91 ± 11.57a[a]	126.53 ± 45.44a
2	0.53 ± 0.29a	35.24 ± 23.15b
3	1.74 ± 0.43a	21.16 ± 8.21b
4	21.64 ± 13.55a	9.39 ± 3.56b
5	6.98 ± 2.63a	5.47 ± 0.88b
F value	NS[b]	—[c]
Fertilizer-amended	3.01×10^1 g⁻¹	1.79×10^1 g⁻¹

Source: Seaker and Sopper (1988a).

Note: Data reflects mean of six samples with standard error.

[a]   Means followed by different letters are significantly different at the 0.05 level of probability by the Waller-Duncan k-ratio t-test.
[b]   NS = no significant effect.
[c]   Significant effect at $P < 0.01$.

compared to numbers of nitrifying bacteria in unamended soils, which have been reported to range from a few hundred to 10^5 g⁻¹ (Stevenson, 1982). This indicates that nitrification was not inhibited on the sludge-amended sites. Because of the continuing release from the organic-N compounds in sludge, the supply of ammonium-N for nitrification remained at a high level (Seaker and Sopper, 1988b). On the fertilizer-amended site, however, *Nitrobacter* reached only 18 g⁻¹, and *Nitrosomonas* only 30 g⁻¹. The addition of N fertilizer to the site did not appear to be sufficient to provide a sustained supply of ammonium-N for a five-year period. The low nitrifying populations suggest a severe lack of organic-N, and partly explains the sparse growth on the site even after five years.

The presence of vegetation on a developing minespoil enhances nitrification (Mills, 1985), as indicated by the high nitrifying bacteria populations on the densely vegetated sludge-amended sites. The pH, however, appears to have a stronger influence than plant cover (Wilson, 1965), with little nitrification occurring below pH 6.0, even on revegetated sites (Jurgensen, 1978). The lower pH value on the five-year old fertilizer-amended site may have contributed to the low level of nitrifying bacteria, compared to the sludge-amended sites that had pH values ranging from 6.7 to 7.3.

Soil Community Respiration

Soil community respiration has long been used as an indicator of biological activity in the soil profile, and be a better estimator of relative microbial activities of minespoils than population counts. Seaker and Sopper (1988a) reported that respiration was significantly higher on a one-year old site (Site 1) than on the older sites due to the "flush" of microbial activity resulting from the readily available organic C addition (Table 79). On sites two to five years old, both the bacterial populations and the community respiration rate declined and stabilized. Production of CO_2 in the minespoils was positively correlated with bacterial populations (r =

Table 79.   Soil Community Respiration and Decomposition Rates on Strip Mine Sites 1 to 5 Years Following Sludge Application, and on Fertilizer-amended Site

Site	CO_2 Evolution (mg CO_2 100_g^{-1} d^{-1})	Decomposition (% yr^{-1})
1	138.08 ± 28.12a[a]	54 ± 8.1c
2	43.92 ± 13.90b	70 ± 5.4b
3	27.22 ± 12.28b	77 ± 1.2b
4	70.66 ± 14.12b	71 ± 2.5b
5	78.69 ± 20.05b	96 ± 0.7a
F value	—[b]	—[c]
Fertilizer-amended	14.17 ± 2.97	—

Source: Seaker and Sopper (1988a).

Note: Data reflects means of three and six samples, respectively, with standard error.

[a]   Means followed by different letters are significantly different at the 0.05 level of probability by the Waller-Duncan k-ratio t-test.
[b,c] Significant effect at $P < 0.01$ and 0.001, respectively.

0.64). Respiration rates on all sludge-amended sites exceeded those on the fertilizer-amended site. Other workers found respiration to be consistently lower in barren minespoils than in conventionally revegetated ones where chemical fertilizer was used, and highest in undisturbed soils (Wilson, 1965; Lawrey, 1977). On conventionally reclaimed spoils amended with chemical fertilizers, respiration increases with the age of the site (Visser, 1985; Stroo and Jencks, 1982), but extensive time periods are required for buildup of microbial populations, often as long as 20 years. On the sludge-amended sites, however, the immediate establishment of microbial communities through organic matter addition appeared to eliminate significant age effects. Establishment of a stable respiration rate appeared to occur within one year, with no significant decrease with site age.

Even the lowest mean respiration rate from the sludge-amended sites, which occurred on the three-year old site, was approximately double that of the five-year old fertilizer-amended site (Seaker and Sopper, 1988a). This indicates that without adequate organic matter input, microbial activity in conventionally reclaimed minespoil remains extremely low.

Microbial Decomposition

Decomposition rate can be used as an indicator of the degree of soil ecosystem recovery, since it largely controls nutrient cycling (Miller and May, 1979). Seaker and Sopper (1988a) reported that on the sludge-amended sites, decomposition rate increased with site age (r = 0.80). After a year of exposure, little more than half the grass sample (54%) was decomposed on Site 1, while almost all of the sample (96%) was decomposed on Site 5 (Table 79). Decomposition was higher on the four older sites, even though microbial numbers were highest on Site 1. This could be attributed to a younger microbial community having less diversity than on the older sites. Site 1 may have a greater proportion of sugar and starch decomposer

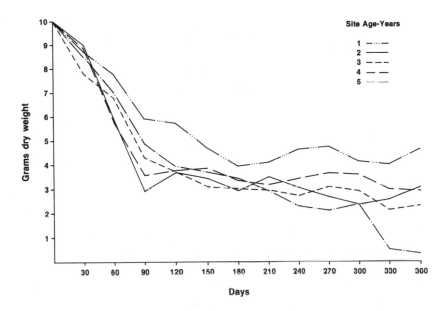

Figure 25.   Amounts of orchardgrass remaining in nylon net bags at 30-day intervals over a one year exposure period for the five sludge-amended sites in Pennsylvania (from Seaker and Sopper, 1988a).

species compared to the older sites that would probably have a greater proportion of species capable of decomposing cellulose and humus. In other studies, where minespoils were reclaimed with chemical fertilizer, populations of cellulose-decomposing fungi and bacteria have been found to increase with time. Younger spoils had a predominance of sugar- and cellulose-decomposers, while older sites had more lignin-degrading species, which resulted in a higher decomposition rate (95%) (Miller and Cameron, 1978; Segal and Mancinelli, 1987).

Figure 25 shows the dry weight loss of the grass samples at 30-day intervals on the sludge-amended sites in Pennsylvania and supports the idea that more mature minespoils possess a wider range of decomposers than do newly reveg-etated ones (Seaker and Sopper 1988a).

Decomposition rates were similar for all sites during the first month, indicating a similar ability to decompose simple organic compounds. After the first 30-day period, Site 1 began to lag behind the other sites. From day 300, the decomposition rate on Site 5 increased significantly compared to Sites 2, 3, and 4. Degradation of the more resistant components of the grass samples was most successfully achieved on the oldest site. Decomposition rate of cellulose has been found to be related to reclamation success, and was significantly increased by vegetative cover (Carrel et al., 1979) and by sewage sludge additions (Parkinson et al., 1980) in other studies. Microbial decomposition improves soil physical conditions by the formation of humus, and the leaching of decomposition products contributes to soil horizonation (Tate, 1985). It is probable that soil formation from minespoil will occur at a faster rate when amended with sludge than when amended with chemical fertilizers.

Trace Metals

Concentrations of HCl-extractable Zn, Cu, Cd, Pb, Ni, and Cr in the sludge-amended minespoils and a fertilizer-amended spoil are given in Table 80 for the Pennsylvania sites. All metals were lowest on Site 1, due to slightly lower metals concentrations in the sludge during that year, combined with a slightly lower application rate. Site 3 had significantly higher Zn, Cu, Cd, and Ni than the other sites, because the soil samples were taken from an area where sludge had been stockpiled prior to spreading. Although the absolute metal loadings on each site varied to some extent, they were all within the maximum allowed by Pennsylvania state regulations, which are very conservative compared to federal guidelines and to metal levels shown to inhibit microbial activities. The surface pH (0 to 15 cm) for Sites 1 through 5 were 6.9, 7.0, 6.8, 6.7, and 7.3, respectively. The pH on the fertilizer-amended site was 6.3. A pH of 6.5 to 8.0 is optimum for the rapid decomposition of wastes in soils, and facilitates the growth of grass-legume forage as well as the immobilization of trace metals in sludge-amended soils.

At high levels, metals such as Zn, Cu, Cd, Pb, and Cr may interfere with microbial functions and have been shown to inhibit soil bacterial and fungal activity (Babich and Stotsky, 1977a). However, investigations in this area have generally utilized solution culture, pot studies, and plate culture techniques rather than land application systems, and purified metal salts rather than sewage sludges. Metal levels evaluated, which ranged from 100 to 10,000 mg kg^{-1} of Zn, Cu, Cd, Pb, Ni, or Cr, were far in excess of those normally encountered in land application systems employing median metal sludges (Babich and Stotsky, 1977b; Premi and Cornfield, 1969; Mathur et al., 1979; Bhuiya and Cornfield, 1974). Conclusions drawn from such studies are not directly applicable to field conditions because effects of metals depend on loading rates, sludge quality, and the form in which each metal occurs, and are strongly influenced by the complexity of the soil system as well as the complexity of the sludge. Binding of metals to humic or fulvic acids, proteins, or crystalline lattices of clay particles, as well as precipitation, complexation, and ionic interactions, influence metal inhibitory effects (Gadd and Driffiths, 1978). Studies have shown that microbial inhibition was reduced when metals were added to soil or organically complexed, and that a diverse soil population containing some tolerant forms would not be significantly affected by metal additions (Martin et al., 1966; Babich and Stotsky, 1977b; Doelman and Haanstra, 1979). Considering the high rates of metals involved in such studies, it is not surprising that with the low range of metal loading rates applied in the Pennsylvania study, microbial populations were within the ranges reported for undisturbed soils.

Very few studies have focused specifically on the effects of metals in land-applied sludge on microbial activity. Tomlinson (1966) suggested that complexing of sludge-borne metals with the components of the sludge/soil mixture may significantly reduce bacterial inhibition by metals. Soil amended with sludge to provide 100, 200, or 400 mg kg^{-1} Zn, Cu, Cd, Pb, and Cr, did not inhibit N transformations except at the highest level (Chang and Broadbent, 1982). Chicago sludge applied to minespoils at rates up to 369 Mg ha^{-1} (dry wt. basis) increased

Table 80.   Extractable Metal Concentrations on Pennsylvania Strip Mine Sites 1 to 5 Years Following Sludge Application, and on Fertilizer-amended Site

Site	Metal Concentrations (mg kg⁻¹)					
	Zn	Cu	Cd	Pb	Ni	Cr
1	32 ± 8.1d[a]	7 ± 3.2c	0.57 ± 0.09d	5 ± 2.1c	3 ± 0.32c	0.9 ± 0.33c
2	89 ± 8.3c	22 ± 2.7b	1.07 ± 0.09c	12 ± 1.3ab	4 ± 0.24c	5.4 ± 0.79ab
3	200 ± 7.7a	36 ± 4.5a	2.77 ± 0.07a	9 ± 1.5bc	11 ± 0.30a	6.1 ± 0.91ab
4	96 ± 4.2c	17 ± 1.8b	1.10 ± 0.06c	15 ± 1.9a	4 ± 1.02c	4.8 ± 0.55b
5	148 ± 10.4b	20 ± 1.1b	1.47 ± 0.22b	13 ± 1.9ab	7 ± 0.64b	7.0 ± 0.38a
F value	—[b]	—[b]	—[b]	—[c]	—[b]	—[b]
Fertilizer-amended	1.3 ± 0.2	1 ± 0	0.17 ± 0.03	3.2 ± 0.2	0.8 ± 0.24	0.3 ± 0.0

Source: Seaker and Sopper (1988a).

Note: Data reflects means of three samples with standard error.

[a] Means followed by different letters are significantly different at the 0.05 level of probability by the Waller-Duncan k-ratio t-test.

[b,c] Significant effect at $P < 0.001$ and 0.05, respectively.

HCl-extractable Zn, Cu, Cd, Pb, Ni, and Cr above typical soil levels, but resulted in no significant reduction in microbial populations, percentage of denitrifiers, or specific enzyme activities (Varanka et al., 1976). Mean metal concentrations in the Chicago sludge were 333 mg kg^{-1} Zn, 70 mg kg^{-1} Cu, 16 mg kg^{-1} Cd, 70 mg kg^{-1} Pb, 11 mg kg^{-1} Ni, and 61 mg kg^{-1} Cr; pH was 6.6 to 6.7. These metal concentrations were considerably higher than on the Pennsylvania sites.

In fact, Zn, Cu, Ni, and Cr applied in sludge have been found in some instances to stimulate microbial activity and plant growth (Varanka et al., 1976; Premi and Cornfield, 1969) due to their role as cofactors for cellular enzyme systems. The fertilizer-amended site was deficient in Zn and Cu, which are essential for plant growth and microbial functioning.

A major problem with revegetating minespoils is the extreme acidity that releases metals such as Zn, Cu, Cd, Pb, Al, and Mn from the spoil into the soil solution at concentrations inhibitory to soil microorganisms (Mills, 1985). Reduced respiration and fungal populations in strip mine spoil have been attributed to high levels of metals coupled with low nutrient levels and acid pH (Lawrey, 1977). In sludge that has been processed by digestion and composting, sludge metals are bound to the organic components as sulfides, chlorides, carbonates, hydroxides, and other compounds not readily soluble. The rapidly established, dense vegetation achieved by sludge amendment would greatly increase water-holding capacity, reducing oxygen infiltration, acid formation, and release of metals from the spoil.

Although minespoils can eventually recover "soil" characteristics through intensive reclamation and management techniques, annual fertilizer additions are usually required for several years. Such methods are rarely practical on vast acreages of nonagricultural land. Without annual maintenance, vegetative cover often deteriorates because microbial development is slow and nutrient cycling never becomes fully operative. The use of sludge as a spoil amendment eliminated the initial lag period that characterizes conventionally reclaimed sites, during which plant growth and microbial activity are at a low level, each one insufficient for maximum functioning of the other. Sludge amendments quickly increased the numbers and activity of microorganisms, whose activities enhance the development of a soil environment conducive to continued plant growth. Development of an indigenous microbial community was achieved on all the Pennsylvania sites, which is a key factor in providing long-term site stability through biogeochemical cycling of energy and nutrients. Recovery of normal soil populations and processes in the surface 5 cm appeared to occur within two years, and did not show a tendency to deteriorate (Seaker and Sopper, 1988a).

Organic Matter

Reestablishing primary production, i.e., a vegetative cover that will persist for the five-year period required by federal surface mining regulations, is the principal aim of reclamation programs. This relatively short-term establishment of a ground cover, however, is not a sufficient guarantee that long-term ecosystem recovery will occur. Soil ecosystem stability results from continuous organic matter accumulation and cycling, and reclamation success may be measured by the degree of change in spoil characteristics toward characteristics indicative of a productive soil. These include detritus accumulation and decomposition; organic matter, organic C, and organic N contents; and root proliferation, all of which are determined by or influenced by soil microbial activity.

Recovery of native organic matter levels, soil structure, and A horizon development may require over 30 years in minespoils from various environments, through natural succession or with conventional reclamation practices using only inorganic fertilizers (Anderson, 1977; Jenny, 1980; Leisman, 1957; Schafer et al., 1980). At best, reclamation with chemical fertilizers requires intensive management and annual fertilizer additions for several years. In practice, however, reclamation efforts are usually minimal, and after initial vegetation establishment, poor physical and biological conditions, not having been addressed, prevent the development of stable C and N cycles. As a result, the plant cover deteriorates before it has a chance to ameliorate the spoil (Stroo and Jencks, 1982; Tate, 1985). For example, even after six years, spent oil shales amended with fertilizer and revegetated did not recover their microbial community, and nutrient cycling and fertility remained low (Segal and Mancinelli, 1987). On fertilizer-amended spoil, organic matter and N levels tended to decrease with age, suggesting an eventual decrease in productivity and site stability after fertilizer amendments are discontinued (Stroo and Jencks, 1982). Nitrogen loss from the site due to poor retention and cycling indicated that site deterioration may not become apparent until after the five-year bonding period required by federal regulations. It is for this reason that ecosystem recovery is seldom achieved with conventional reclamation techniques.

Organic amendments can be extremely important to successful reclamation (Sopper and Seaker, 1984; Down and Stocks, 1977; Visser, 1985). Numerous studies conducted over the past 60 years have shown that amending soils with organic materials such as sewage sludge increases soil organic matter content, and improves soil structure and long-term fertility (Joost et al., 1987). Organic matter decomposition and cycling, processes difficult to initiate in disturbed soils, are quickly achieved by sludge additions. Depending upon the amount of topsoil material present, it is possible on some sites to achieve significant soil development in minespoils in as few as three to five years through intensive management and the use of amendments that accelerate plant growth and soil-forming processes (Visser, 1985).

The reclamation success achieved with sludge is due to three factors related to its organic content:

1. The N content is in a slowly available organic form.
2. The high organic C content provides an immediate energy source for soil microbes.
3. Sludge organic matter improves the poor spoil physical conditions resulting from soil removal and compaction.

In the Pennsylvania study, previously cited, Seaker and Sopper (1988b) also reported on the effects of the sludge applications on dry matter yield, detritus accumulation, root growth, and spoil organic matter on the five sites ranging in age from one to five years after sludge application.

Table 81.   Dry Matter Yield and Detritus Accumulation on Strip Mine Sites 1 to 5 Years Following Sludge Application, and on Fertilizer-amended Site

Site	Dry Matter Yield (Mg ha⁻¹)	Detritus Accumulation (g m⁻²)
1	4.1 ± 0.47 c[a]	96 ± 11d
2	13.5 ± 0.99ab	264 ± 51c
3	18.6 ± 1.78a	954 ± 45a
4	11.1 ± 0.52b	533 ± 14b
5	16.6 ± 3.76a	370 ± 60c
F values	—[b]	—[b]
Fertilizer-amended	5.2 ± 0.9	286 ± 50

Source: Seaker and Sopper (1988b).

Note: Data reflects mean of six samples with standard error.

[a]   Means followed by different letters are significantly different at the 0.05 level of probability by the Waller-Duncan k-ratio t-test.
[b]   Sigificant effect at $P < 0.001$.

The dry matter yields of vegetation on the sludge-amended and control site are given in Table 81. The biomass produced on Sites 2 through 5 was more than double that of the five-year old site amended with chemical fertilizer. Yield was significantly higher on Sites 2 through 5 than on Site 1, and was correlated with site age ($r = 0.48$). By the third year, a maximum yield of 18.6 Mg ha⁻¹ was reached. Forage produced on all sludge-amended sites compared favorably with maximum hay yields from agricultural land in the same counties, which were 5.0 and 4.2 Mg ha⁻¹ in 1981, respectively (USDA, 1981). Yield was also correlated with soil organic matter ($r = 0.65$), which indicates that the increasing plant biomass C gradually replaced the sludge C compounds as a source of microbial energy.

A dense vegetative cover restricts movement of water downward in the spoil which results in decreased O_2 infiltration and reduces formation of acid by sulfuroxidizing bacteria. It can also improve and buffer temperature and moisture extremes, which, in turn, enhances N and C mineralization and nitrification (Tate, 1985; Mills, 1985). The more rapid establishment and greater density of plant cover achieved by sludge amendments is beneficial for soil development from minespoil because it enhances the cycling of C and N.

The accumulation of dead organic material on the surface of Sites 2 through 5 was greater than that of Site 1 and the fertilizer-amended site (Table 81). Maximum accumulation occurred on Site 3. Detritus was positively correlated with dry matter yield ($r = 0.48$), indicating that nutrients were being recycled through the decomposition of vegetation. The five-year old control site had a detritus accumulation similar to that of Site 2, and considerably less than the older sludge-amended sites. On sites reclaimed with short-term chemical fertilizer applications, growth is usually too sparse and too slow to provide a substantial layer of detritus.

The decomposition of dead organic material and its subsequent incorporation into the spoil is an indication that soil rejuvenation is in progress. The protective organic cover moderates the high surface temperatures of the spoil environment,

Table 82.   Root Biomass Determined from Whole Plants on Strip Mine Sites 1 to 5 Years Following Sludge Application, and on Fertilizer-Amended Site

	Biomass (g plant^{-1})		
Site	Birdsfoot Trefoil	Orchardgrass	Composite
1	0.6 ± 0.4c[a]	5.0 ± 0.8c	2.7 ± 0.8b
2	15.0 ± 3.9b	15.7 ± 3.4bc	15.4 ± 2.4b
3	14.7 ± 2.2b	56.4 ± 10.9a	35.5 ± 8.2a
4	30.9 ± 3.5a	40.0 ± 18.1ab	35.5 ± 8.9a
5	25.3 ± 7.8ab	48.7 ± 14.1ab	37.0 ± 8.5a
F values	—[b]	—[c]	—[b]
Fertilizer-amended	17.4 ± 4.0	25.2 ± 3.7	21.3 ± 2.8

Source: Seaker and Sopper (1988b).

Note: Data reflects mean of six samples with standard error.

[a]   Means followed by different letters are significantly different at the 0.05 level of probability by the Waller-Duncan *k*-ratio *t*-test.

[b,c]   Sigificant effect at $P <0.001$ and 0.05, respectively.

promotes water retention and seed germination, decreases soil loss by erosion, and, by leaching into the spoil, contributes to the formation of soil horizons. Since microbial activity is strongly influenced by litter, the return of microbial communities to minespoil would be expected to occur more rapidly on sites with a more productive grass cover. When sludge was applied to the surface of coarse coal refuse, a densely rooted organic matter layer up to 7 cm deep developed within two years (Scanlon et al., 1973); and on subalpine and prairie surface mine sites, sludge additions increased litter input compared to sites reclaimed with either chemical fertilizer or peat (Visser, 1985).

Composite root mass on Sites 3 through 5 was greater than that of the fertilizer-amended site (Table 82). Greatest orchardgrass root growth was observed on Sites 3 through 5, and birdsfoot trefoil root growth on Sites 4 and 5. Compared to the five-year old fertilizer-amended site, the root systems on Site 5 were 45 to 93% larger. Root biomass of orchardgrass, birdsfoot trefoil, or both combined, was positively correlated with site age ($r = 0.49$, 0.65, and 0.48, respectively). Orchardgrass roots were correlated with soil organic matter, organic C, and N ($r = 0.65$, 0.80, and 0.67, respectively). This suggests that the higher organic matter content of the sludge-amended spoils enhanced root proliferation by decreasing bulk density and increasing microbially activated nutrient cycling. Restricted rooting often occurs on conventionally reclaimed spoils low in organic matter if clays have been compacted or if drought occurs in coarse-textured material (Smith et al., 1971). This was evident in the root systems of the fertilizer-amended site that tended to be shallow and saucer-shaped, while those on the sludge sites were deeper. Root exudates and decaying root cells largely determine the microbial composition of the soil environment, and the rhizosphere of grasses is particularly important in minespoil where soil properties are unfavorable for microbial growth in the surrounding spoil (Cundell, 1977). This may account for significant reductions in microbial populations in barren spoils compared to

Table 83.   Organic Matter Content, C and N Content, and C/N Ratio of Strip Mine
             Spoil 1 to 5 Years Following Sludge Application, and of Fertilizer-
             Amended Site

Site	% Organic matter	% Organic C	% Kjeldahl N	C/N Ratio
1	1.60 ± 0.09c[a]	3.09 ± 0.44b	0.37 ± 0.06	8
2	2.93 ± 0.25b	6.13 ± 1.38ab	0.41 ± 0.10	15
3	5.13 ± 0.30a	6.69 ± 0.35ab	0.41 ± 0.08	17
4	3.72 ± 0.40b	3.80 ± 0.46b	0.30 ± 0.07	14
5	4.52 ± 0.34a	7.59 ± 1.72a	0.48 ± 0.12	16
F value	—[c]	—[b]	NS[b]	NS
Fertilizer-amended	1.75 ± 0.38	1.57 ± 0.23	0.10 ± 0.01	15

Source: Seaker and Sopper (1988b).

Note: Data reflects mean of six samples for organic matter, and three samples for C and N, with standard error.

[a]   Means followed by different letters are significantly different at the 0.05 level of probability by the Waller-Duncan k-ratio t-test.
[b,c] Significant effect at $P <0.05$ and 0.001, respectively.
[d]   NS = no significant effect.

revegetated ones. At approximately the same pH, a barren mine site was reported to have 0.07×10^6 bacteria g^{-1}, while a revegetated site had 74×10^6 bacteria g^{-1} in the rhizosphere of the grass cover (Wilson, 1965).

The organic matter content of the spoil from sites 2 through 5 was greater than on the fertilizer-amended site (Table 83). A more dense vegetative cover and higher decomposition rates (Seaker and Sopper, 1988b) on sites 2 through 5 increased the organic matter content up to threefold compared to the fertilizer-amended site. The older sludge sites had significantly higher organic matter content than Site 1, and organic matter was correlated with site age ($r = 0.67$). The maximum was observed on Site 3. Similar organic matter increases with site age have been reported for subalpine minespoils (Visser et al., 1983), and 369 Mg ha⁻¹ (dry wt. basis) of sludge increased the organic matter content of silt loam minespoil to 5.43% (Varanka et al., 1976).

Soil horizon development partially depends on such microbial processes as vegetation decomposition, incorporation of humus into the spoil, downward movement of organic material in the developing profile, and organic sorting (Ollier, 1969). Such processes are enhanced by the addition of sludge to minespoil and by the subsequent increased biomass production compared to spoils amended with chemical fertilizers. The organic matter content of the fertilizer-amended site remained at a low level, suggesting that the initial stages of soil horizon develop-ment were inhibited.

There was no difference in N content between the sludge-amended sites, which indicates that N levels established by sludge application were maintained even on the five-year old site (Table 83). Nitrogenous organic additions, e.g., proteins contained in sludge, result in net mineralization of N that builds up the N pool. Because it is mainly in the organic form, the N in sludge becomes slowly available over time. Much of the N in this system is conserved and recycled. Losses of N due to nitrate leaching or ammonia volatilization were not measured but appear to be insignificant in comparison to the amount of N retained in the spoil.

An N level of 0.25% has been reported for undisturbed soils in the eastern USA (Wilson, 1965; Stevenson, 1982). The sludge-amended sites had N contents of 0.30 to 0.48%, while the fertilizer-amended site had an N content of 0.10%. This indicates that reclamation with sludge can restore and even improve soil fertility with respect to N for a period of time beyond that achieved with chemical fertilizers. Nitrogen accumulation is difficult to achieve with inorganic N fertilizers. Stroo and Jencks (1982) found minespoils reclaimed with fertilizer and mulch initially productive, but little of the N remained in the soil, suggesting that self-sustaining ecosystems are not being achieved through current reclamation practices.

Seaker and Sopper (1988b) reported that organic C content was increased at least threefold on the sludge-amended sites compared to the fertilizer-amended site, and ranged from 3 to 7% (Table 83). In contrast, the C content of 20-year old strip mine sites, reclaimed without an organic amendment, were reported to average only 0.93% (Wilson, 1965). Organic C was lowest on Site 1 and highest on Site 5, but there was no significant correlation with site age. Chemical fertilizers can provide initial stimulation of plant growth, but unlike organic amendments, they do not directly stimulate microbial activity by providing a C source.

The higher values for organic C than for organic matter may be due to the difference in methodology. In the Walkley-Black method, organic matter is oxidized by chromic acid, while the Coulometric method is a dry combustion procedure. It is possible that the Coulometric method oxidized a higher percentage of the C present in soil-sized particles of coal, which have been reported to be a source of error in the determination of organic matter in minespoil (Wilson, 1965; Schaefer et al., 1980). Despite the differences in organic C content achieved by the two methods, both sets of data follow the same general trend, with the lowest values for the fertilizer-amended spoil, and the highest values for sites three and five years old.

Seaker and Sopper (1988b) found that the C/N ratio approximately doubled after the first year due to accumulation of plant biomass (Table 83), and appears to have stabilized at about 16 by the second year. This ratio is between that of nutritionally balanced agricultural soils, which is about 25 according to Stevenson (1982), and that of most native soils, which is 10 to 12 (Reeder and Berg, 1977). The C/N ratio of the fertilizer-amended site was 15. However, the absolute amounts of C and N were very low compared to the sludge-amended sites, reflecting the low organic matter content and poor fertility that would largely account for the inferior plant growth. Both C and N were increased nearly fivefold on the sludge-amended sites compared to the control. There was no evidence of deterioration of either organic C or N in the sludge-amended spoils over time. Similarly, Hinesly et al. (1979) reported that residual C and N applied in sludge at a rate of 61 Mg ha[-1] remained stable for four years after the sludge application stopped.

In a companion study, Sopper and Seaker (1988) compared microbial community development and activity on an abandoned strip-mine spoil bank partially

Table 84. Average Concentrations of Constituents in Sludge Collected at Time of Application

Constituent	Mean Concentration (dry wt.) (mg kg⁻¹)
NO³–N	205
NH⁴–N	3,331
Org–N	7,718
Total N	11,254
Total P	10,199
K	1,928
Ca	14,224
Mg	3,801
Na	705
Fe	29,715
Al	16,473
Mn	863
Zn	1,435
Cu	579
Cr	332
Pb	628
Ni	62
Co	17
Cd	5
Hg	1
pH	8.3
Solids (%)	48

Source: Sopper and Seaker (1988).

amended with sludge at 128 Mg ha⁻¹ in Clearfield County, Pennsylvania. Six years after sludge application, soil samples were collected from the sludge-treated plot and the surrounding barren area and analyzed for microorganism populations and activity by standard methods. The sludge product applied was Philadelphia mine mix, which is a mixture of dewatered sludge cake and composted sludge cake (50:50 mixture by volume). The average concentrations of constituents in the

Table 85. Comparison Between EPA and PDER Interim Guidelines for Maximum Trace Element Loading Rates for Land Reclamation and Amount Applied

Element	EPA (CEC 5-15) (kg ha⁻¹)	Maximum Loading Rate, PDER (kg ha⁻¹)	Amount Applied at 128 Mg ha⁻¹ (kg ha⁻¹)
Zn	560	224	184
Cu	280	112	74
Cr	Na	112	43
Pb	800	112	81
Ni	280	22	8
Cd	11	3	0.7
Hg	Na	0.6	0.1

Source: Sopper and Seaker (1988).

a No recommendations given by EPA.

Table 86.   Comparison of Microbial Populations on the Sludged and Barren Sites

Site	Aerobic Heterotrophic Bacteria 10^6 g^{-1}	Fungi 10^5 g^{-1}	Actinomycetes 10^4 g^{-1}
Sludge	6.64 ± 2.52	1.60 ± 0.38	5.88 ± 2.43
Barren	0.13 ± 0.25	0.001 ± 0.00008	0.01 ± 0.009
t-test	a	b	a

Source: Sopper and Seaker (1988).

[a] $P <0.05$.
[b] $P <0.01$.

sludge collected at the time of application are given in Table 84. Amounts of trace metals applied in comparison to regulatory recommendations are given in Table 85.

The populations of aerobic heterotrophic bacteria, fungi, and actinomycetes are given in Table 86. All microbial populations were significantly higher on the sludge-amended plot. Municipal sewage sludge itself contains numerous species of bacterial and fungi, some of them pathogenic, as well as actinomycetes. Most of the bacteria found in sludge are of human or animal original while others have the specific function of decomposing organic matter during anaerobic digestion and composting. This inoculum is of little importance, however, compared to the indigenous soil microbial populations stimulated in the spoil by the addition of sludge organic matter (Fresquez and Lindemann, 1982). Providing a carbon source is more important than providing an inoculum source for stimulating an active and varied spoil microflora. Intestinal microorganisms in sludge are quickly destroyed in and on the soil by dessication, sunlight, and predation by microorganisms adapted to the soil environment.

The bacteria populations on the sludge-amended site (6.64×10^6 g^{-1}) are higher than that reported for a spent oil shale site in Colorado that had been fertilized, revegetated, and weathered for six years (1.5×10^6 g^{-1}) (Segal and Mancinelli, 1987). Estimates from the literature of "normal" bacterial populations in undisturbed soils generally range from 1 to 34×10^6 per gram. Thus, it appears that the sludge application has facilitated the development of a healthy bacteria population. Similarly, the fungal population on the sludge-amended site (1.6×10^5 g^{-1}) is within the "normal" range of 0.05 to 9.0×10^5 g^{-1} reported in the literature. The actinomycete population on the sludge-amended site (5.88×10^4 g^{-1}) is clearly on the low side of what has been reported in the literature as "normal" populations for undisturbed soils, 1 to 436×10^4 per gram.

Similarly, populations of nitrifying bacteria were significantly higher on the sludge-amended plot (Table 87). Populations of both genera are low but still are within the normal range for undisturbed soils, which have been reported to range from a few hundred to 10^5 per gram (Stevenson, 1982). *Nitrosomonas* and *Nitrobacter* were virtually nonexistent on the barren portion of the site. Spoil pH is reported to have a strong influence on nitrifying bacteria populations, with little nitrification occurring below pH 6.0 even on revegetated sites (Jurgensen, 1978).

Table 87.   Comparison of Nitrifying Bacteria Populations on the Sludged and Barren Sites

Site	Nitrosomonas 10^4 g^{-1}	Nitrobacter 10^5 g^{-1}
Sludged	1.52500 ± 0.63279	3.169085 ± 0.674916
Barren	0.00004 ± 0.00002	0.000006 ± 0.000001
t-test	a	b

Source: Sopper and Seaker (1988).

[a] $P < 0.05$.
[b] $P < 0.01$.

At the time of this study average spoil pH was 5.0 on the sludge-amended site and 4.1 on the barren site.

Soil community respiration has long been used as an indicator of biological activity in the soil profile, and may be a better estimator of relative microbial activities between soils than population counts. Results of laboratory measurements of CO_2 evolution are given in Table 88. Soil respiration was significantly higher on the sludge-amended site and was correlated with the higher bacterial populations. The rate of CO_2 evolution on the sludge-amended site was similar to that reported for another site three years after being amended with the same sludge at the same rate (28 vs 27 Mg CO_2 100 g^{-1} day^{-1}) (Sopper and Seaker, 1987a).

Table 88.   Comparison of CO_2 Evolution Measurements Made on Sludged and Barren Sites

Site	Lab mg CO_2 100 g^{-1} day^{-1}
Sludged	28.18 ± 7.50
Barren	1.61 ± 0.12
t-test	a

Source: Sopper and Seaker (1988).

[a] $P < 0.05$.

Table 89.   Comparison of Decomposition on the Sludged and Barren Sites

Site	Organic Matter Decomposed in 1 Year
Sludged	74.7 ± 2.6
Barren	50.3 ± 1.8
t-test	a

Source: Sopper and Seaker (1988).

[a] $P < 0.01$.

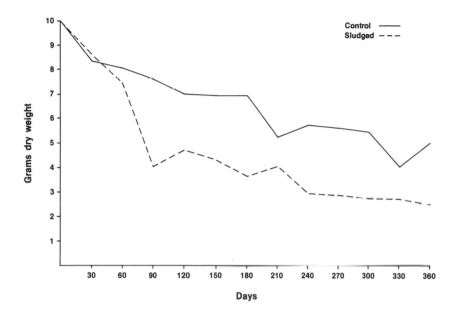

Figure 26.   Amounts of orchardgrass remaining in nylon net bags at 30-day intervals over a one-year exposure period.

Monthly and annual decomposition rates of organic matter in the litter bags are given in Table 89 and Figure 26. Organic matter decomposition was significantly higher on the sludge-amended site. After one year of exposure 75% of the orchardgrass was decomposed on the sludge-amended site in comparison to 50% on the barren site. Decomposition rates were similar during the first 30 days (April 15 to May 15). After that period, the decomposition rate greatly accelerated on the sludge-amended site. In a similar study, average annual organic matter decomposition rates ranged from 54 to 96% for sites that were one to five years old after being amended with the same sludge at approximately the same application rate (Sopper and Seaker, 1987b).

The significance of microbial decomposition to site recovery is threefold:

1. It improves spoil physical conditions by the formation of humus.
2. It indicates nutrient and energy cycling.
3. The decomposition products leach downward, contributing to the formation of soil horizons.

Since the speed with which the ecological community recovers depends largely on the amount of organic matter reaching the soil, an increasing decomposition rate is indicative of rapid ecosystems recovery.

Spoil organic matter content was significantly higher (P <0.001) on the sludge-amended site (5.00 ± .43%) than on the barren site (1.35 ± 0.10%). This was probably due to the organic matter additions in the sludge and the dense vegetative cover providing a greater amount of detritus for incorporation into the surface spoil. Soil ecosystem stability results from continuous organic matter accumulation and cycling and reclamation success may be measured by the degree of change in spoil characteristics toward characteristics indicative of a productive soil. All of these processes are influenced by soil microbial activity.

Nitrogen Mineralization

Lack of plant-available N may be a major problem in revegetation of some lands disturbed by surface mining. Sewage sludges contain 20 to 60g N kg^{-1} or more, much of which is in the organic form (Sommers et al., 1976). The rate of mineralization of this organic N to NH_4–N and the subsequent nitrification to NO_3–N is important in supplying adequate N for revegetation establishment.

Disagreement exists among researchers concerning the effect of the rate of sewage sludge addition on the percent of added organic N mineralized. Terry et al. (1981) using sewage sludge application rates of 11.2, 22.4, and 44.8 Mg ha^{-1} found that the percent of added organic N mineralized was significantly greater at higher rates than the lower rates of sludge addition. Epstein et al. (1978) and Magdoff and Chromec (1977) observed no rate effect on the percent of added organic N mineralized at rates ranging from about 20 to 80 Mg ha^{-1}. Following this trend, Sabey et al. (1977) observed that the percent of added organic N mineralized decreased as the amount of N added increased.

Voos and Sabey (1987) conducted a 16-week laboratory incubation study to determine the rate of net N mineralization in sewage-sludge-amended coal mine spoil. Sewage sludge was added at rates of 0, 40, 80, and 120 Mg ha^{-1} which added 0, 1630, 3260, and 4890 kg N ha^{-1}, respectively. The total amount of inorganic N that accumulated during the experiments increased significantly as the rate of sewage sludge addition increased. Only small amounts of NO_3–N had accumulated in the mine spoil after 16 weeks. Ammonium-N increased with increasing rates of sewage sludge. The total net N mineralized in 16 weeks in the mine spoil treated with 0, 40, 80, and 120 Mg ha^{-1} of sludge were 274, 402, 505, and 617 mg kg^{-1}, respectively.

Hornby et al. (1986) also studied nitrogen mineralization potentials of lignite overburden in Texas. Nitrogen mineralization potentials were determined from laboratory data on a premined native soil and four-year old reclaimed mixed overburden that received 180 kg N ha^{-1} yr^{-1} as $NH_4 NO_3$. In addition, the effect of anaerobically digested and dewatered sewage sludge on N$^-$ mineralization potential of overburden was evaluated on samples (0 to 15 cm) collected from field plots. Samples were collected at 2, 26, and 52 weeks after amendment. Treatments were 0 kg N ha^{-1}, 212 kg N ha^{-1} as $NH_4 NO_3$, 106 kg N ha^{-1} as NH_4NO_3 plus 106 kg N ha^{-1} as sludge, and 212 kg N ha^{-1} as sludge. Samples were incubated in the lab for 18 weeks. The N mineralization potentials for all sewage sludge plots remained higher than all other treatments at the end of one year. Release of NO_3–N was significantly higher in plots that received sewage sludge than in the control or fertilized plots. The total N in sludge-amended plots did not change significantly with time over the 52-week period. They concluded that a single application of sludge at 56 Mg ha^{-1} to overburden provided a greater supply of mineralizable N than that resulting from four years of mineral N fertilizer applications.

Earthworms

Sludge applications to agricultural soils and mine spoils can have a significant effect on earthworm populations. Earthworm activity is important in soil formation processes (Miller, 1974) and are useful indicators of soil metal availability (Van Hook, 1974). On sludge-amended soils, Hartenstein et al. (1981) reported that earthworms accumulate high concentrations of metals. This may pose a potential hazard to the earthworms and their predators. Earthworm populations are usually very low or nonexistent on mine spoils, thus, few studies have been conducted.

A search of the literature resulted in finding only one study by Pietz et al. (1984) that was conducted at the Fulton County, Illinois land reclamation site. Anaerobically digested sewage sludge has been applied continually since 1972 to calcareous strip mine spoil (Peterson et al., 1982). The three-year (1975 to 1977) study sampled mine soil and nonmined fields to determine the effect of land application of anaerobically digested sludge on the heavy metal accumulations in earthworms. The only earthworm species found on the mine soil fields was *Aporrectodea tuberculata*. On the nonmined fields the species *Lumbricus terrestris* was also found. Sewage sludge applications to fields on both land types (mine soil vs nonmined) resulted in significant accumulations of Cd, Cu, and Zn. Land type significantly affected earthworm Zn concentrations, with levels being higher in all nonmined fields sampled. Earthworm Cd and Cu accumulations in all fields sampled were significantly related to the current amounts of sludge-applied metals and the amount applied since the previous sampling. Concentrations of Ni, Cr, and Pb in earthworms were not significantly related to sewage sludge applications during the 1975 to 1977 sampling period. Earthworm metal concentrations in all fields studied ranged from 19 to 506 mg Zn kg^{-1}, 0.6 to 98 mg Cd kg^{-1}, 0.1 to 8.8 mg Cr kg^{-1}, 1.1 to 25 mg Cu kg^{-1}, <0.1 to 10 mg Ni kg^{-1}, and <0.1 to 2.4 mg Pb kg^{-1}. The higher values all occurred on sludge-amended fields. Little is known about the toxicities of soil metals to earthworms. Hartenstein et al. (1981) determined the concentrations at which heavy metals added to activated sludge would induce a toxic effect on the growth of *Eisenia foetida*. All of the maximum metal concentrations cited above were from sludge-amended soils containing metal levels considerably lower than the threshold toxicity levels listed by Hartenstein et al. (1981). Since Cu, Cd, and Zn were accumulated in significant amounts by worms in sludge-amended soils at the Fulton County site, there may be a potential hazard to predators (primarily birds) with continued long-term sludge applications.

Summary

Sludge applications on mine land generally enhances the reestablishment of functioning microbial communities. Near normal functioning communities of bacteria, fungi, and actinomycetes are usually established within two to three years following sludge applications. Sludge amendments quickly increased the number and activity of microorganisms and eliminates the initial lag period that characterizes conventionally reclaimed mine sites. The quick establishment of microbial populations increases organic matter decomposition and results in rapid recycling of nutrients on the site which benefits vegetation. Although high concentrations of trace metal may inhibit microbial activity, most research results have shown that trace metals applied in sludge have not had an adverse effect on microbial populations or activity.

EFFECTS ON WATER QUALITY

Some concern has been voiced over the effects of high rates of sludge application on the quality of groundwater and nearby streams, ponds, and lakes. The state of Pennsylvania, for example, prohibits the use of sludge for land reclamation directly on a watershed area that supplies drinking water to a community. But that is not to say that sludge application results in the deterioration of waters. In fact, the opposite is usually true. Stabilization of drastically disturbed lands with municipal sludge often improves the quality of the surrounding area in having an ameliorative effect on the ecosystem as a whole. Reports on the effects of sludge application on concentrations of NO_3–N, trace metals, and on indicator organisms in soil percolate water, groundwater, nearby streams and lakes, and surface runoff indicate that a properly managed land application program will not cause deterioration of water quality on or near the site.

Soil Water and Groundwater

Nitrate–Nitrogen

There is a potential for nitrate buildup and eventual leaching into groundwater when sludge, particularly liquid sludge, is applied continuously (Kardos et al., 1979; Urie et al., 1982). NO_3–N of percolate water has increased in varying degrees after sludge application (Kardos et al., 1979; Urie et al., 1982; Sopper and Kerr, 1981). Irrigating with liquid sludge and sewage effluent at 2.5 and 5.0 cm per week, totaling 60 and 120 cm on bituminous spoil, and 75 and 150 cm on anthracite refuse, increased nitrate and ammonium in the leachate water. But NO_3–N did not exceed the drinking water limit of 10 mg L^{-1} (Kardos et al., 1979). Nitrates below the drinking water limit were also reported for leachate 90 cm below the surface of a burned anthracite refuse bank amended with sludge at 40 to 150 Mg ha^{-1} (Kerr et al., 1979). Some studies show an initial peak in NO_3–N concentration followed by a decrease to acceptable levels. Often, peak NO_3–N levels occur in late winter and spring when plants are not utilizing the nitrogen for growth (Haghiri and Sutton, 1982). Both Cl and nitrate levels in leachate from sludge-amended sand and gravel spoils were initially increased, but tended to decrease over an 8 week period (Hornick, 1982). The NO_3–N concentration of groundwater collected from wells at various depths on sludge-amended mine spoils in Pennsylvania was consistently within safe drinking water standards, when monitored monthly for up to five years (Sopper and Kerr, 1981).

Table 90.   **Mean Annual Concentrations of NO₃–N and Trace Metals in Groundwater Collected on Strip Mine Site in Pennsylvania Amended with 184 Mg ha⁻¹ Sludge**

Site	Year[a]	pH	NO₃-N	Cu	Zn	Cr	Pb	Cd	Ni
					Concentrations (mg L⁻¹)				
Well 1	1977	4.4	1.4	0.22	4.13	0.02	0.14	0.006	3.67
(Control)	1978	4.3	<0.5	0.23	2.02	0.01	0.19	0.002	0.98
	1979	4.6	<0.5	0.17	1.48	0.03	0.13	0.001	0.50
	1980	5.5	0.6	0.05	0.89	0.05	0.09	0.001	0.50
	1981	5.7	0.7	0.06	0.83	0.03	0.04	0.003	0.31
	1989[b]	5.8	0.02	0.01	0.09	<0.001	0.01	0.001	0.06
Well 2	1977	4.6	1.1	0.10	3.39	0.03	0.09	0.001	2.67
(Sludge)	1978	4.5	<0.5	0.14	3.29	0.01	0.20	0.002	1.26
	1979	4.4	<0.5	0.18	1.49	0.03	0.13	0.001	0.97
	1980	5.7	0.6	0.05	1.05	0.04	0.11	0.001	0.76
	1981	6.0	0.6	0.05	0.57	0.02	0.05	0.001	0.31
	1989[b]	6.6	0.06	0.01	0.07	<0.001	0.01	0.001	0.04
EPA Drinking Water Standard			10	1	5	0.05	0.05	0.010	—

Source: Sopper and Seaker (1990).

[a] Values are annual means of monthly samples.
[b] Average of three samples collected in August 1989.

Seaker and Sopper (1984) reported that an application of sludge of 184 Mg ha⁻¹ on an abandoned strip mine had little effect on groundwater quality, even though the water table was only 3 to 4 m below the surface. They reported that average monthly concentrations of NO₃–N were below 10 mg L⁻¹ (maximum concentration for potable water) for all months during a five-year period. The highest monthly value recorded was 2.4 mg L⁻¹. Groundwater samples were also analyzed for total and fecal coliforms. No fecal coliform colonies were observed for any monthly sample. Sopper and Seaker (1990) resampled groundwater in 1989, 12 years after sludge application, and found NO₃–N concentrations still at an extremely low level (0.06 mg L⁻¹) (Table 90).

The application of lime and sludge and subsequent revegetation appears to have had a positive effect on groundwater pH. Groundwater pH increased from 4.6 (1977) to 6.0 by 1981. Results of the 1989 sampling indicates a pH of 6.6. There has also been a gradual increase in pH in the control well from pH 4.4 to pH 5.8. Since 1980, attempts were being made to reclaim the control area by conventional methods using lime and fertilizer. The amounts of lime and fertilizer applied and frequency of application are not known as the coal company is no longer in business. However, these applications and vegetation growth probably contributed to the increase in groundwater pH in the control well.

There appears to be no significant increase in any of the trace metal concentrations over the initial five-year period (1977 to 1981) in the groundwater samples from Well 2 compared to the control well (Table 90). From 1977 to 1981 most of the monthly concentrations were within the U.S. EPA drinking water standards. The only exception was Pb which exceeded the limit of 0.05 mg L⁻¹ for both the control well and Well 2, probably resulting from increased release of the element from the spoil material due to mining. The highest monthly Pb values were 0.28 mg L^{-1} in the control well and 0.33 mg L^{-1} in Well 2 in 1978, and the mean annual Pb concentrations were 0.19 and 0.20 mg L^{-1} for control well and Well 2, respectively. By 1981, however, the mean annual Pb concentrations had decreased to 0.04 and 0.05 mg L^{-1} for the two wells. Results of analyses of the groundwater samples collected in 1989 given in Table 90 show that all trace metal concentrations in both wells were extremely low in comparison to values for the initial five years (1977 to 1981).

Similarly, on an anthracite coal refuse bank treated with 80 and 108 Mg ha^{-1} of sludge, Seaker and Sopper (1983) found little effect on groundwater quality. Over a five-year monitoring period all monthly values of NO$_3$–N were less than 10 mg L^{-1}. Groundwater samples were also analyzed for total and fecal coliforms and no fecal coliforms were ever observed for any sample.

Pietz et al. (1989c) reported on the effects of applying sludge (542 Mg ha^{-1}), lime (89.6 Mg ha^{-1}), and gypsum (112 Mg ha^{-1}) and various combinations to coal refuse on percolate water quality at a depth of 1 m. Samples were collected monthly over a five-year period. Yearly mean concentrations of NH$_4$-N, [NO$_3$ + NO$_2$]–N ranged from 0.8 to 225, and 0.0 to 278 mg L^{-1}, respectively. The high values were all associated with the sludge treatment. The NH$_4$–N in percolate from the sludge-amended treatments was initially high, but declined with time to near background levels after five years. The [NO$_3$ + NO$_2$]–N concentrations in the sludge-amended treatments increased the first three years reaching a maximum of 278 mg L^{-1}, and then declined rapidly in subsequent years to levels of 20 to 61 mg L^{-1} by the fifth year.

Trace Metals

On anthracite refuse and bituminous strip-mine spoils, leachate collected 107 cm below a grass, legume, and tree seedling cover was lower in Fe, Al, and Mn where liquid sludge and sewage effluent were applied biweekly than it was in control spoils (Kardos et al., 1979). Satisfactory renovation of the major constituents of sludge through acidic strip mine spoil has been reported by McCormick and Borden (1973). The pH of the percolate water was related to the sludge rate and application method. Initially high sulfur levels resulting from the sludge decreased to below those in control areas. In Ohio, strip-mine spoils amended with sludge at rates up to 716 Mg ha^{-1}, leachate concentrations of Cu, Ni, and Mn did not increase, and even decreased with time. Zinc and Al initially increased, but then decreased, while Cd and Pb were below detection limits (Haghiri and Sutton, 1982). On a burned anthracite refuse bank, sludge applications of 75 to 150 Mg ha^{-1} did not degrade the quality of percolate water 90 cm below the surface (Kerr et al., 1979). In fact, Zn and Cd concentrations were lower in the sludge treated plots than in the control plots. Where sludge applications were monitored for three to five years on three 4-ha sites in Pennsylvania, groundwater samples collected monthly showed no evidence of contamination (Sopper and Kerr, 1981; Sopper and Seaker, 1982, 1983, 1984). Copper, Zn, Cr, and Pb were, with very few

Table 91.   Mean Annual Concentrations of pH, Nitrate-N and Trace Metals in Groundwater

	Year	pH	NO$_3$-N	Cu	Zn	Cr	Pb	Co	Cd	Ni
					Concentrations (mg L^{-1})					
Well 1 (Control) Well 2	1978[a]	6.3[b]	<0.5	0.22	1.69	0.02	0.18	0.18	0.002	0.23
(108 Mg ha^{-1})	1978	6.8	<0.5	0.13	0.51	0.02	0.08	0.02	<0.001	0.01
	1979	6.8	<0.5	0.10	0.56	0.01	0.04	0.01	<0.001	0.03
	1980	7.3	0.8	0.05	0.48	0.01	0.03	0.01	<0.001	0.03
	1981	6.8	2.2	0.03	0.04	0.03	0.02	0.01	<0.001	0.01
	1982	6.7	2.4	0.01	0.11	0.02	0.04	0.01	<0.001	0.01
EPA Drinking Water Standard			10	1.0	5.0	0.05	0.05	5.0[c]	0.01	2.0[c]

Source: Sopper (1991a).

[a] Well vandalized in 1978.
[b] Values are annual means of monthly samples.
[c] Recommended values for irrigation water for agricultural use.

exceptions, well within safe drinking water standards established by the U.S. EPA. Lead sometimes exceeded the limit by a minimal amount, even on unsludged areas, because of Pb-bearing minerals in the bedrock.

Applications of municipal sludge at 80 and 108 Mg ha^{-1} on an anthracite refuse bank had little effect on trace metal concentrations in groundwater (Sopper, 1991). Results of analyses of groundwater well samples over a five-year period are given in Table 91. The values for Well 1 (control) reflect quality of groundwater for the disturbed mine site. Well 2 reflects the effects of dewatered sludge on water quality. Results indicate that the sludge applications have not had any significant effect on groundwater pH or NO$_3$–N. Average monthly effect on concentrations of NO$_3$–N were below 10 mg L^{-1} for all months sampled during the five-year period.

There was a marked decrease in the trace metal concentrations over the five-year period in the groundwater samples from Well 2 compared to the control well. From 1978 to 1982 the monthly concentrations of Cu, Zn, Cr, and Cd were within the U.S. Environmental Protection Agency drinking water standards. The only exception was Pb which occasionally exceeded the limit of 0.05 mg L^{-1} for both the control well and Well 2, probably resulting from increased release of the element from the refuse material due to regrading the bank. The highest monthly Pb values were 0.20 mg L^{-1} in the control well and 0.10 mg L^{-1} in Well 2 in 1978. The mean annual Pb concentrations were 0.18 and 0.08 mg L^{-1} for control well and Well 2, respectively. By 1982 the mean annual Pb concentration of Well 2 had decreased to 0.04 mg L^{-1}. No groundwater samples could be collected in 1990 as all wells had been vandalized in the interim period.

In an ongoing program, over 1500 ha of strip-mined land in Pennsylvania have been reclaimed with Philadelphia sludge applied at 134 Mg ha^{-1}. Over a five-year period, groundwater quality met drinking water standards for metals and fecal coliform bacteria (Sopper and Kerr, 1980a,b; Sopper 1982a–e; Sopper et al., 1981).

Pietz et al. (1989c), in the same study as previously cited, found that the concentrations of all metals in the percolate at 1 m, except Pb and Hg, were significantly affected by the sludge, lime, and gypsum treatments. Aluminum, Fe, Cu, Ni, Cd, and Zn levels were generally lowest in the sludge or control treatments. However, concentrations of these same metals increased with time, indicating a solubilization of these metals with time in both the sludge and sludge + lime treatments. The authors concluded that the applications of sludge (542 Mg ha^{-1}) and lime (89.6 Mg ha^{-1}) based on theoretical calculations to control acidity were too low for long-term reclamation. The authors recommend that for long-term reclamation (>5 yr) of coal refuse, an application of lime or sewage sludge alone should be 189 and 1050 Mg ha^{-1}, respectively. If both materials are used, lime and sewage sludge application rates between 134 to 189 and 900 to 1350 Mg ha^{-1}, respectively, would be desirable for reclamation on a long-term basis.

The U.S. Army Corps of Engineers (1987) reported on the successful use of digested dewatered municipal sewage sludge to revegetate dredged and excavated material from the Chesapeake and Delaware Canal which had been deposited into diked areas along the canal. These dredged materials are acidic and low in organic matter and nutrients and subject to erosion by wind and water. Municipal sewage sludge from the city of Baltimore, Maryland was used as the amendment. Sites treated were located in both Maryland and Delaware and included a total of 253 ha. Sludge was applied, after liming to raise the dredged material to pH 7.0, and then disced. Sludge application rate was 168 Mg ha^{-1} on the Delaware site and 112 Mg ha^{-1} on the Maryland site in accordance with State regulations. The sites were seeded with a mixture of perennial grasses consisting of Kentucky bluegrass (18.6%), tall fescue (59.8%), red fescue (18.6%), and weeping lovegrass (3.0%) and then mulched. Trace metal concentrations in the Black River Wastewater Treatment plant sludge applied were 23, 1002, 351, 5, 233, and 2607 for Cd, Cu, Pb, Hg, Ni, and Zn, respectively. Pre- and post-water sample analyses data collected over a three-year period from 40 monitoring wells were combined and statistically analyzed. Depth to water table ranged from 3 to 21 m in the groundwater monitoring wells. There was no statistically significant differences in concentrations of Cd, Ni, Cu, and Zn between pre- and post-samples. Concentrations of Pb in post-samples were significantly lower. Prior to sludge application, none of the groundwater parameters sampled fell within the Federally established primary and secondary maximum contaminant levels for drinking water. After sludge application all trace metal concentrations were within the Federal drinking water limits. Mean NO$_3$–N concentrations were not significantly different between pre- and post-sludge samples (0.084 vs 0.080 mg L^{-1}).

Soil samples were collected from the 20 to 30 cm soil depth before and after sludge application and combined for statistical analyses. Soil concentrations of Cd, Co, Cr, Cu, Hg, Ni, Pb, and Zn increased significantly as a result of sludge application. However, the metal concentrations fell within naturally occurring soil ranges for these elements as reported by Shacklette and Boergen (1984). In comparison with Allaway's (1968) range of naturally occurring soil elemental levels, only Cu exceeded the natural range (127 mg kg^{-1} vs 2 to 100 mg kg^{-1}).

Vegetation foliar samples were collected from 99 plots on the sludge treated sites and analyzed for trace metals. Of the trace metals analyzed (As, B, Ba, Cd,

Cu, Fe, Pb, Mn, Ni, and Zn), only Ni and Mn exceeded their respective naturally occurring ranges (Allaway, 1968) and only Ni (8.53 mg kg^{-1}) exceeded its suggested tolerance level (3 mg kg^{-1}) as reported by Blakleslee (1978). Even though Ni levels were relatively high, it is not considered to be phytotoxic below 50 mg kg^{-1} (Melsted, 1973). Although Mn (256 mg kg^{-1}) exceeded the naturally occurring concentration range, it fell well below the suggested tolerance level (300 mg kg^{-1}). It was concluded that no plant tissue concentration was high enough to pose a serious hazard to either vegetation or animals.

Surface Water

Surface water runoff and a stream adjacent to the Palzo site were monitored following applications of liquid sludge (Urie et al., 1982; Jones and Cunningham, 1979). Reduction of surface runoff was related directly to the density of the vegetative cover. NH_4–N, NO_3–N, and total N concentrations were decreased in runoff analyzed two years after sludge application, compared to runoff from unsludged areas (Urie et al., 1982). Iron, SO_4, Al, and Cd concentrations in a nearby stream were drastically increased as a result of strip mining, but reductions in ion concentrations during sludge application did not degrade water quality during the period (Jones and Cunningham, 1979). The chemical and biological quality of Contrary Creek, adjacent to an abandoned pyrite mine in Virginia, was not affected within a year after the site was revegetated using municipal sludge. Because of runoff from barren areas and the toxic mine sediments already in the stream bed, it may take decades before site stabilization affects stream quality (Hinkle, 1982). On the Fulton County project in Illinois, the water-quality monitoring program calls for monthly sampling of 33 wells and nearby streams and reservoirs. Although sludge addition increased nitrate N minimally, mean annual concentrations of NO_3–N, Cd, Zn, Cu, Cr, and Pb were within U.S. EPA drinking water limits in the two watershed reservoirs tested during the two years after sludge was applied. Fecal coliform counts were not increased by sludge application; in fact, they decreased, probably because fewer livestock grazed on the site. The authors (Peterson et al., 1979) concluded that a properly managed, digested sludge application site will not adversely affect local surface waters.

Two lakes adjacent to an abandoned bituminous strip mine in Pennsylvania were monitored monthly for five years, after the site was reclaimed with liquid and dewatered sludges at rates up to 184 Mg ha^{-1}. For the entire five-year period, nitrate-N, Zn, Cu, Cr, and Cd were below the U.S. EPA maximum limits for drinking water. Nitrates were slightly increased the first several months after sludge application, but then decreased. Lead was often slightly above drinking water limits due to natural dissolution of Pb bearing minerals in the underlying rock (Sopper and Kerr, 1981).

In West Virginia, Skousen (1988) reported on the effects of surface applications of sludge, without incorporation, on water quality in three ponds which received the drainage from the entire 6.5 ha area treated. Sludge applications were 22.4, 44.8, and 78.4 Mg ha^{-1}. Water samples were collected prior to sludge application and at two and six month intervals after application. Analyses of water samples for pH, NO_3–N, Cr, Zn, Cu, Cd, Ni, Pb, and Al showed few differences before and after sludge applications. NO_3–N concentrations were 3.7 mg L^{-1} before sludge application and 3.0 and 2.0 mg L^{-1} for the two post application sampling dates.

Recently, Daniels and Haering (1990) reported on the application of Philadelphia mine mix (a 50:50 mixture of wood chip compost and sludge cake) at a design rate of 112 Mg ha^{-1} to an active surface mine site in Virginia. Actual application was 100 Mg ha^{-1} of sludge along with 200 Mg ha^{-1} of composted wood chips. After incorporation by chisel plowing and disking, the site was broadcast seeded with a hayland/pasture seed mix and winter rye. Water samples collected from a discharge from a hollow fill monthly for four months after sludge application indicated nitrate concentrations ranging from 1.88 to 2.34 mg L^{-1} in comparison to 4.35 mg L^{-1} prior to sludge application. Water samples collected from a large sediment pond receiving surface drainage from the site also only ranged from 1.20 to 2.19 mg L^{-1} after sludge application in comparison to 1.92 mg L^{-1} for presludge samples.

Summary

Stabilization of drastically disturbed mine land with municipal sludge often improves the quality of the site and surrounding area in having an ameliorative effect on the ecosystem as a whole. Research results on the effects of sludge applications on the concentrations of NO_3–N, trace metals, and on indicator organisms (fecal coliforms) in soil percolate, groundwater, nearby streams and lakes, and surface runoff indicate that a properly managed land application program will not cause deterioration of water quality on the site.

EFFECTS ON ANIMAL NUTRITION AND HEALTH

The quality of forage should be determined before livestock are grazed on lands reclaimed with municipal sludge, as it would be in the management of a normal farming operation. In a properly managed land application system, forages would be expected to be of good nutritional quality. On five (4-ha) sites in Pennsylvania where mine spoils were revegetated with several types of sludge at rates from (7 to 202 Mg ha^{-1}), tall fescue, orchardgrass, and birdsfoot trefoil were of excellent quality. Trace metal concentrations were low, and protein and fiber contents were comparable to those in forages grown on agricultural land amended with inorganic fertilizers (Seaker and Sopper, 1982). Sludge amendments on mine spoils in Illinois significantly improved corn grain quality as measured by protein content (Blessin and Garcia, 1979). Two studies showed some potential problems that may be encountered. Nitrogen:sulfur ratios were low in tall fescue (5:1) grown in West Virginia strip mine spoils amended with up to 224 Mg ha^{-1} of sludge, as compared with recommended ratios of 10:1 to 15:1 (Mathias et al., 1979). On soil contaminated by a nearby zinc smelter, forages grown with municipal sludge were not considered suitable for feed because of high nitrates, Zn, and Cd the first two growing seasons (Franks et al., 1982).

Plant uptake of metals and sludge deposits on leaves eaten by grazing animals can increase the potential for higher tissue concentrations. Proper use of sludge depends on the impact on soils, plants, and animals exposed to it. According to Fitzgerald (1982), animals exposed to excess heavy metals show toxic reactions rather quickly. In a study involving the Fulton County, Illinois project the concentrations of seven trace metals in tissues from animals grazing in sludge-grown forage were not significantly different from those in the controls, except for increased Cd, Pb, Cu, and Zn in the liver and Cd in the kidney. Lead levels in the blood were increased fourfold. The diaphragm, heart, brain, bone, and milk showed no trace of sludge, nor was there any effect on the reproductive rate, or any evidence of disease. The growth of experimental cows was above average. In a similar study, pigs feeding in sludge-amended pens did not accumulate any more trace metals in the diaphragm, heart, or bone than did control animals, but Cd was increased in liver and kidney tissue.

Another study with Chicago sludge assessed metal accumulations in various organs of pheasants and swine which were fed for 100 and 56 days, respectively, corn grain harvested from reclaimed areas where liquid sludge had been applied annually for five to six years. Annual loading rates ranged from 25 to 128 Mg ha^{-1}. Because trace metal composition of muscle tissue was unaffected by a diet of sludge-grown corn, the authors concluded that the consumption of meat from these animals would present little, if any, potential health hazard to humans. Although Cd concentrations were increased significantly in the liver and kidney tissues of pheasants and swine fed the corn grain from the sludged fields as compared with concentrations in control animals, the maximum levels of Cd in the tissues were still comparable to those reported elsewhere for animals fed a normal diet (Hinesly et al., 1979b).

When red-winged blackbirds nesting on the sludge-reclaimed Palzo strip mine were analyzed for tissue Cd, Zn, and Pb, it was found that Pb concentrations in their brain, liver, kidney, and muscle were no different than those in birds living in undisturbed areas or on strip mine sites reclaimed with inorganic fertilizers (Gaffney and Ellertson, 1979). Higher kidney Cd was observed on the sludged sites, but in some tissues, Cd tissue concentrations were higher in birds from natural areas than in birds living on the Palzo site. Interpretation of such data is complicated due to Cd–Zn interactions that may occur in animal tissues and to the lack of a data base for normal metal levels in birds.

In Pennsylvania, two recent studies were reported which investigated the trace metal concentrations in the tissue of cottontail rabbits (*Sylvilagus floridanus*) and meadow voles (*Microtus pennsylvanicus*) trapped on an abandoned strip mine site revegetated using Philadelphia sludge applied at 134 Mg ha^{-1}. The goal of both studies was to follow the transfer of trace metals from sludge to the soil and into the animals food chain (Dressler et al., 1986 and Aberici et al., 1989). The amounts of trace metals applied in the sludge were 175 kg Zn ha^{-1}, 53 kg Pb ha^{-1}, 50 kg Cu ha^{-1}, 31 kg Cr ha^{-1}, 7 kg Ni ha^{-1}, and 2 kg Cd ha^{-1}. Extractable trace metal concentrations in surface soil samples (0 to 15 cm) were higher on the sludge-amended site but were not significantly ($P > 0.05$) different from a control site reclaimed by conventional methods using lime and chemical fertilizer (Table 92). Vegetation species on both sites were similar. Concentrations of Zn in all plant species and Cd and Cu in three of the four plant species were higher ($P < 0.05$) on the sludge-amended site compared to the control site (Table 93).

Table 92.   Extractable Trace Metal Concentrations (Dry Weight) in Surface Soil Samples

	Control (mg kg⁻¹)		Treated (mg kg⁻¹)		
Metal	Mean	SE	Mean	SE	Normal Range in Soil[a]
Cu	2.92±	0.16	6.68±	2.47	2–100
Zn	1.50±	0.24	78.91±	23.41	10–300
Cr	0.24±	0.03	2.04±	0.91	5–3000
Pb	2.67±	0.15	3.53±	0.88	2–200
Co	1.37±	0.10	1.00±	0.15	1–40
Cd	0.06±	0.00	0.39±	0.09	0.01–7.0
Ni	2.21±	0.31	3.11±	0.77	10–1000

Source: Dressler et al. (1986).

[a] Allaway (1968).

In the first study, ten adult rabbits were trapped on the treated site and eleven adult rabbits were trapped on the control site (Dressler et al., 1986). Trace metal concentrations found in rabbit tissues are given in Table 94. Levels of most metals in cottontail rabbits collected from both sites were not significantly different between males and females. Concentrations of Zn in cottontail rabbit femurs on the treated site were higher ($P < 0.05$) than those on the control site, corresponding to increased Zn concentrations in the vegetation on the treated site. Concentrations of Cu, Cr, Pb, Co, Cd and Ni in all tissues (femur, kidney, liver, and muscle) were not significantly different between the two sites.

In a companion study, 36 adult female rabbits were held in laboratory cages and fed a diet with additions of cadmium sulfate hydrate ($3CdSO_4 \cdot 8H_2O$), a highly soluble Cd compound. Treatments consisted of a control (Purina Lab Rabbit Chow HF5326), control with 5 mg Cd kg⁻¹, and control with 25 mg Cd kg⁻¹. The rabbits were necropsied after 126 days on the diet. Addition of the Cd salt to the diets resulted in an increase ($P < 0.01$) in Cd in all cottontail rabbit tissues except muscle (Table 95). As expected, highest levels of Cd were found in the kidneys and liver and Cd levels increased with increased dietary intake of Cd.

Laboratory data using metal salts must be used cautiously and cannot be directly compared to field data where Cd is added in a sludge product. Logan and Chaney (1983) pointed out the many problems of comparing plant uptake of soluble metal salts with metals in digested sludge and comparing controlled laboratory studies with field experimental results. If soluble metal salts are taken up at a greater rate than sludge metals (Logan and Chaney, 1983), any field values above the laboratory levels might indicate a potential problem. Levels of Cd in rabbit tissue samples in this study from the mined sites (control and sludge-amended) and from nonmined sites, except muscle and liver tissues from the sludge-treated site, were comparable to or below ($P < 0.05$) the laboratory control levels. Cadmium concentrations in muscle from the sludge-treated mine site were higher than the 5 mg Cd kg⁻¹ laboratory treatment group. Although the muscle Cd concentrations from mined and nonmined sites were higher ($P < 0.05$) than laboratory controls, there was no significant difference between the sites, and these levels were below mean levels reported in selected foods analyzed by the Food and Drug

Table 93.   Heavy Metal Concentrations (Dry Weight) in Vegetation Foliar Samples[a]

Element	Control (mg kg^{-1})			Treatment (mg kg^{-1})		
	Mean	SD	Median	Mean	SD	Median
			Bromus sp.			
Cu	5.8	0.7	5.5[c]	8.4	2.1	7.4[c]
Zn	21.9	2.1	21.5[c]	41.8	13.9	37.6[c]
Cr	0.0	0.0	0.0	0.6	1.5	0.0
Pb	2.7	1.4	2.9	2.0	0.9	1.9
Co	0.3	0.4	0.1	<0.1	0.1	0.0
Cd	0.02	0.02	0.01	0.06	0.07	0.03
Ni	0.5	0.4	0.5	1.6	2.9	0.6
			Orchardgrass			
Cu	9.4	0.8	9.4	10.5	1.4	10.4
Zn	23.7	2.0	23.5[c]	43.9	7.0	45.5[c]
Cr	0.0	0.0	0.0[c]	8.6	17.4	1.3[c]
Pb	4.2	0.7	4.0	3.2	1.1	3.3
Co	0.5	0.3	0.5	0.2	0.2	0.1
Cd	0.02	0.02	<0.01[c]	0.2	0.07	0.2[c]
Ni	1.0	0.6	1.0[b]	4.7	6.2	2.3[b]
			Tall fescue			
Cu	7.0	0.6	7.0[b]	8.0	0.5	7.9[b]
Zn	21.7	1.4	21.8[c]	41.4	8.8	37.6[c]
Cr	0.1	0.3	0.0	4.2	10.2	0.0
Pb	2.8	0.8	2.0	2.3	1.1	2.1
Co	0.1	0.2	0.0	0.3	0.4	0.0
Cd	0.02	0.02	0.02[c]	0.1	0.06	0.12[c]
Ni	1.4	0.9	1.4	4.4	2.4	1.1
			Trefoil			
Cu	10.7	0.4	10.8[b]	11.6	0.7	11.5[b]
Zn	43.0	1.5	43.0[b]	60.0	2.8	59.9[b]
Cr	0.3	0.6	0.0[b]	20.5	15.3	25.1[b]
Pb	4.3	0.4	4.5	3.6	0.3	3.7
Co	1.8	0.0	1.8[b]	0.5	0.5	0.4[b]
Cd	0.02	0.02	0.03[b]	0.2	0.04	0.16[b]
Ni	13.8	0.6	13.5	17.0	4.7	18.4

Source: Dressler et al. (1986).

[a]   All values based on six samples except control-trefoil (*n*=3).
[b,c] Significant at $P \le 0.05$ and $P \le 0.01$, respectively.

Administration, e.g., ground beef mean = 0.075 mg Cd kg^{-1} (Sharma, 1981). All Cd concentrations in muscle tissues from rabbits collected on the control and sludge-treated mined sites were below the corresponding values reported by Curnow et al. (1977) in cottontails collected on southeastern Ohio watersheds. Cadmium levels in livers from the sludge-treated site (median = 2.25 mg kg^{-1} were higher than those from the laboratory control group; however, no difference (P <0.05) was noted when these were compared with cottontail rabbit livers collected from a non-mined

Table 94.   Heavy Metal Concentrations (Dry Weight) of Male and Female Cottontail Tissues Collected in March 1983[a]

Element	Control (mg kg⁻¹)			Treatment (mg kg⁻¹)		
	Mean	SD	Median	Mean	SD	Median
Femur						
Cu	2.9	0.3	3.0	2.8	0.2	2.8
Zn	126	13	125[b]	148	22	147[b]
Cr	1.4	0.2	1.5	1.3	0.4	1.3
Pb	12.2	1.1	12.0	12.9	1.0	12.8
Co	5.3	0.5	5.5	5.4	0.6	5.4
Cd	0.01	0.01	0.01	0.01	0.01	0.01
Ni	6.8	0.6	6.8	6.5	0.4	6.3
Kidney						
Cu	10.6	2.6	11.0	10.9	1.2	11.3
Zn	81.6	23.9	77.0	87.5	26.3	84.3
Cr	0.0	0.0	0.0	0.0	0.0	0.0
Pb	1.5	1.2	1.5	2.3	2.4	1.5
Co	0.0	0.0	0.0	0.1	0.2	0.0
Cd	9.6	7.1	5.3	17.0	14.1	12.3
Ni	0.5	0.8	0.0	0.8	1.0	0.0
Liver						
Cu	11.0	1.9	10.8	11.8	2.2	11.4
Zn	110	20	120	115	26	114
Cr	0.0	0.0	0.0	0.0	0.0	0.0
Pb	0.2	0.5	0.0	0.8	0.7	0.9
Co	0.0	0.1	0.0	0.2	0.2	0.0
Cd	1.60	0.79	1.63	2.40	0.85	2.25
Ni	0.1	0.1	0.0	0.2	0.2	0.3
Muscle						
Cu	4.3	0.5	4.0	4.0	0.5	4.3
Zn	41.1	3.7	41.8	44.1	5.5	43.6
Cr	0.1	0.4	0.0	0.5	1.7	0.0
Pb	0.3	0.4	0.0	0.1	0.3	0.0
Co	0.1	0.2	0.0	0.0	0.1	0.0
Cd	0.01	0.01	0.01	0.01	0.01	0.01
Ni	0.5	1.0	0.3	0.9	2.0	0.1

Source: Dressler et al. (1986).

[a]   Control values based on 11 replicates; treatment values based on 10 replicates (replicate = one sample from one rabbit).
[b]   Significant at $P < 0.05$

area. Mean levels of Cd in rabbit livers from the sludge-treated site were above mean Cd amounts for selected foods evaluated by the FDA, e.g., raw beef liver mean = 0.183 mg kg⁻¹, but within the range of Cd values observed for several foods, e.g., ground beef, breakfast cereal, and sugar (Sharma, 1981). Authors concluded that the occasional consumption of cottontail rabbit muscle tissue from the sludge-treated site should pose no threat to human health.

Table 95.   Cadmium Concentrations (Dry Weight) of Tissues Collected from Female Cottontails Fed Experimental Diets for 126 Days[a]

Tissue	Control (mg kg⁻¹)			5 mg Cd kg⁻¹			25 mg Cd kg⁻¹		
	Mean	SD	Median	Mean	SD	Median	Mean	SD	Median
Femur	1.0	<0.1	1.0a[b]	1.3	0.4	1.2b	1.7	0.4	1.6c
Kidney	16	14	9a	98	15	95b	341	76	355c
Liver	1.3	0.9	0.9a	13.5	4.1	11.5b	59.6	19.2	57.5c
Muscle	0.00	0.00	0.00a	<0.01	0.01	0.00a	0.19	0.11	0.16b

Source: Dresster et al. (1986).

[a]   Control and 5 mg Cd kg⁻¹ values based on nine replicates, 25 mg Cd kg⁻¹ values based on 12 replicates (replicate = one sample from one rabbit).
[b]   All tests of treatment effects were significant ($P < 0.01$); significant pairwise comparisons are indicated by different letters ($P < 0.01$).

In the second study conducted by Alberici et al. (1989), meadow voles were collected with snap traps and separated by sex and age. Liver, kidney, muscle, and bone tissue from 8 to 10 voles were pooled according to site, sex, and tissue for trace metal analyses. Concentrations of Cu, Zn, Co, Cd, and Ni in meadow vole tissues were not significantly different between the control and sludge-amended site (Table 96). However, Cr concentrations in kidney and bone, and Pb concentrations in liver and bone were significantly higher ($P \leq 0.05$) on the control site than on the sludge-amended site. The highest concentrations of Zn, Cd, Ni, Pb, and Co were found in bone tissue, whereas Cu tended to accumulate in the kidney. This is in contrast to a study by Johnson et al. (1978), who found that Cd tended to accumulate in the kidney. However, Anderson et al. (1982) demonstrated that Cd can accumulate differentially on the basis of age and sex. All the meadow voles in the Pennsylvania study were adults and Cd showed no differential accumulation between sexes.

Zinc, Cd, and Pb concentrations in vole tissues in the Pennsylvania study were lower than those found in meadow voles in a variety of polluted and nonpolluted areas studied by Johnson et al. (1978). Copper, Zn, Pb, and Cd levels in vole tissues in the Pennsylvania study were comparable to those found in meadow voles in a sludge-treated field studied by Anderson et al. (1982). Both Anderson et al. (1982) and Johnson et al. (1978) found no short-term toxic effects on the meadow vole as a result of the trace metals found in their studies. Long-term effects of trace accumulation in the meadow vole are not well documented. In the Pennsylvania study, there was no clear pattern evident of the movement of trace metals from the sludge to the soil and vegetation and finally to the meadow vole.

In addition to trace metal accumulation, another health-related subject of much concern is that of pathogenic organisms in sludge. There is considerable information available on the survival of pathogens in sludge and sludge-amended soils, but little on the occurrence of disease transmission to animals from organisms in sludge (Fitzgerald, 1979). With anaerobically digested and with composted sludge especially, pathogens have not been found to pose a serious health risk. Very few pathogenic organisms survive anaerobic digestion at 35 to 38°C, and even fewer

Table 96.   Extractable Trace Metal Concentrations (Dry Weight) in Meadow Vole Tissues

		Control (mg kg⁻¹)		Treated (mg kg⁻¹)	
Tissue	Metal	Mean	SE	Mean	SE
Kidney	Cu	11.70±	0.15	12.29±	0.33
	Zn	55.07±	1.24	59.32±	0.31
	Cr[a]	0.68±	0.08	0.51±	0.07
	Pb	1.06±	0.26	2.27±	0.32
	Co	0.72±	0.07	0.77±	0.06
	Cd	0.48±	0.02	1.41±	0.22
	Ni	1.01±	0.10	0.59±	0.11
Liver	Cu	13.36±	0.20	13.60±	0.13
	Zn	81.26±	1.18	83.27±	0.88
	Cr	0.43±	0.30	0.37±	0.25
	Pb[a]	2.50±	0.20	0.41±	0.11
	Co	0.49±	0.04	0.40±	0.03
	Cd	0.23±	0.03	0.27±	0.04
	Ni	0.30±	0.14	0.21±	0.15
Muscle	Cu	7.50±	0.28	7.03±	0.10
	Zn	45.35±	0.77	49.16±	0.92
	Cr	0.43±	0.39	0.00±	0.00
	Pb	3.29±	0.13	3.00±	0.07
	Co	0.86±	0.06	0.91±	0.04
	Cd	0.38±	0.03	0.30±	0.01
	Ni	2.49±	0.35	2.38±	0.06
Bone	Cu	4.26±	0.53	3.09±	0.18
	Zn	179.28±	3.09	157.65±	2.72
	Cr[a]	4.53±	1.26	0.35±	0.16
	Pb[b]	12.90±	0.40	11.53±	0.42
	Co	5.77±	0.19	5.15±	0.23
	Cd	1.93±	0.08	1.47±	0.09
	Ni	8.80±	0.64	8.27±	0.35

Source: Alberici et al. (1989).

[a,b] Significant at *P* ≤0.05 and *P* ≤0.01, respectively.

survive composting at temperatures above 55°C. However, with aerobically digested or raw sludge, there may be problems. A four-year study related to the Fulton County, Illinois project examined 100 cows that were grazed on anaerobically digested sludge-treated forage (Fitzgerald, 1979). No bacterial, viral, or fungal infections were observed in live animals or in blood or tissues at necropsy. No tissue parasites were found, and the incidence of intestinal parasites was the same in experimental and control animals. In a similar four-month swine study, *Ascaris* species infected some of the pigs in the pens amended with 200 Mg ha⁻¹ of digested sludge, but the number of worms was small. No other parasites or disease organisms were found, indicating that the transfer of pathogens from anaerobically digested sludge to grazing animals is "remote."

On the 6,000 ha Fulton County project, land application of Chicago sludge resulted in no significant public health problems (Sedita et al. 1977). Actual cases of disease on such projects are extremely scarce, and problems can be eliminated with proper sludge processing and application. Four years of data on the project indicated no significant numbers of viruses or indicator organisms in groundwater or surface water, no differences in discharge of nematode eggs or coccidian oocysts from animals grazing on sludge-treated forage, and no significant difference in soil pathogen content from control and sludged areas. Health effects of digested sludge are quite different from those resulting from the application of raw sludge or of raw or treated wastewater effluents.

Summary

Sludge applications on mine land generally have not had an adverse effect on the health of domestic or wild animals and birds.

U.S. EPA 40 CFR Part 503 Final Rules For Use and Disposal of Sewage Sludge

Under the Clean Water Act of 1972, amended in 1987 by Congress, the U.S. EPA was required to develop regulations for the use and disposal of sewage sludge. The Congressional mandate set forth a comprehensive program for reducing the potential environmental risks and maximizing the beneficial use of sludge. This required the EPA to assess the potential for pollutants in sewage sludge to affect public health and the environment through a number of different routes of exposure. Evaluation of the risks posed by pollutants that may be present in sludge applied on land required the Agency to consider human exposure through inhalation, direct ingestion of soil fertilized with sewage sludge and through consumption of crops grown on this soil. The Agency also had to assess the potential risk to human health through contamination of drinking water sources or surface water when sludge is disposed on land. The Final Rules were signed by the EPA Administrator and were published (Federal Register, 1993). An advanced copy of the Final Rules has been provided by the Water Environment Federation (Water Environment Federation, 1993).

This entire book has summarized research on the use of municipal sewage sludge for reclamation of mine land under the previous EPA regulations and guidelines. The following information on the new regulations is provided to give the reader the option to reinterpret any conclusions drawn by the author.

The numerical criteria in the Final Rule are given in Table 97. These rules state that sewage sludge shall not be applied to land if the concentration of any pollutant in the sludge exceeds the ceiling concentration shown in Table 97. In addition, the cumulative loading rate for each pollutant shall not exceed the cumulative pollutant loading rate nor should the concentration of each pollutant in the sludge exceed the concentration for the pollutant (APL) as shown in Table 97. The annual pollutant loading rate is also given in Table 97. This annual pollutant loading rate generally applies to applications of sewage sludge on agricultural lands.

The Rule also states that the amount of nitrogen applied by sewage sludge on agricultural land should not exceed the nutrient needs of the vegetation grown on the site. This however, does not necessarily apply to land reclamation applications. For land reclamation, the sewage sludge is usually only applied once at a high rate to provide a sufficient pool of nutrients to sustain the vegetation for 3 to 5 years to assure a permanent vegetative cover. The Rule allows the permitting authority to authorize a variance from this requirement provided the operator of the site can demonstrate that the higher nitrogen application in excess of vegetation requirements would not contaminate ground or surface water.

The Rule also includes some management practices which must be followed. Sludge cannot be applied to a reclamation site that is flooded, frozen, or snow-covered if there is the possibility that pollutants might enter a wetland or other waters of the United States. There are other management practices related to pathogens and vector attraction reduction. There are numerous alternatives to meet these requirements and they will not be discussed here. The reader is advised to refer to the Final Rule when published in the Federal Register in February 1993.

Table 97. Numerical Criteria for 40 CFR Part 503 Final Rule

Element	Concentration Ceiling mg kg^{-1}	Cumulative Load kg ha^{-1}	APL[a] mg kg^{-1}	APLR[b] kg ha yr^{-1}
Arsenic	75	41	41	2.0
Cadmium	85	39	39	1.9
Chromium	3000	3000	1200	150
Copper	4300	1500	1500	75
Lead	840	300	300	15
Mercury	57	17	17	0.85
Molybdenum	75	18	18	0.90
Nickel	420	420	420	21
Selenium	100	100	36	5
Zinc	7500	2800	2800	140

[a] Alternate pollutant limits or "clean sludge."
[b] Annual pollutant loading rate.

CHAPTER 3

Conclusion

The use of municipal sewage sludge in reclamation and revegetation of drastically disturbed land has been extensively investigated. The results to date are encouraging and show that stabilized municipal sludges, if applied properly according to present guidelines, can be used to revegetate mined lands in an environmentally safe manner with no major adverse effects on the vegetation, soil, or groundwater quality and do not pose any significant threat to animal or human health. Revegetation of coal strip-mine spoils, gravel spoils, coal refuse, clay strip-mine spoils, iron-ore tailings, abandoned pyrite mine spoils, and sites devastated by toxic fumes has been demonstrated by numerous studies using a variety of types and application rates of sludge. Results from many of the studies have substantiated that the present state and federal guidelines provide adequate protection of the environment as well as animal and human health.

CHAPTER 4

References

Adriano, D. C. *Trace Elements in the Terrestrial Environment* (New York: Springer, 1986) p. 533.

Alberici, T. M., W. E. Sopper, G. L. Storm, and R. H. Yahner. "Trace metals in soil, vegetation, and voles from mine land treated with sewage sludge," *J. Environ. Qual.* 18:115–120 (1989).

Alexander, M. *Introduction to soil microbiology* (New York: John Wiley & Sons, 1977).

Allaway, W. H. "Agronomic controls over the environmental cycling of trace metals," *Adv. Agron.* 20:235–271 (1968).

Anderson, D. W. "Early stages of soil formation on glacial till mine spoils in a semi-arid climate," *Geoderma* 19:11–19 (1977).

Anderson, T. J., G. W. Barrett, C. S. Clark, V. J. Elia, and V. A. Majeti. "Metal concentrations in tissues of meadow voles from sewage-sludge-treated fields," *J. Environ. Qual.* 11:272–277 (1982).

Babich, H. and G. Stotsky. "Sensitivity of various bacteria, including actinomycetes, and fungi to cadmium and the influence of pH on sensitivity," *Appl. Environ. Microbiol.* 33:681–688 (1977a).

Babich, H. and G. Stotsky. "Effect of cadmium on fungi and on interactions between fungi and bacteria in soil: Influence of clay minerals and pH," *Appl. Environ. Microbiol.* 33:1059–1066 (1977b).

Baker, D. E. and L. Chesnin. "Chemical monitoring of soils for environmental quality and animal and human health," *Adv. Agron.* 27:305–374 (1975).

Baker, D. E., D. R. Bouldin, H. A. Elliott, and J. R. Miller, Eds. "Criteria and Recommendations for Land Application of Sludges in the Northeast," Bulletin 851, Pennsylvania Agricultural Experiment Station, The Pennsylvania State University, University Park, PA (1985).

Bastian, R. K., A. Montague, and T. Numbers. "The potential for using municipal wastewater and sludge in land reclamation and biomass production as an I/A technology: an overview," in *Land Reclamation and Biomass Production with Municipal Wastewater and Sludge,* W. E. Sopper, E. M. Seaker, and R. K. Bastian, Eds. (University Park, PA: The Pennsylvania State University Press, 1982), pp. 13–54.

Bayes, C. D., C. M. A. Taylor, and A. J. Moffat. "Sewage sludge utilization in forestry: the U.K. research program," in *Alternative Uses for Sewage Sludge,* J. E. Hall, Ed. (Oxford, U.K.: Pergamon Press, 1990), pp 115–138.

Berry, C. R. "Sewage sludge and reclamation of disturbed forest land in the Southeast," in *Land Reclamation and Biomass Production with Municipal Wastewater and Sludge,* W. E. Sopper, E. M. Seaker, and R. K. Bastian, Eds. (University Park, PA: The Pennsylvania State University Press, 1982), pp. 317–320.

Bhuiya, M. R. and A. H. Cornfield. "Incubation study of effect of pH on nitrogen mineralization and nitrification in soils treated with 1000 ppm lead and zinc oxides," *Environ. Pollut.* 7:161–164 (1974).

Blessin, C. W. and W. J. Garcia. "Heavy metals in the food chain by translocation to crops grown on sludge-treated strip mine land," in *Utilization of Municipal Sewage Effluent and Sludge on Forest and Disturbed Land,* W. E. Sopper and S. N. Kerr, Eds. (University Park, PA: The Pennsylvania State University Press, 1979), pp. 471–482.

Boesch, M. J. "Reclaiming the strip mines at Palzo," *Compost Sci.* 15(1):24–25 (1974).

Carrel, J. E., K. Wieder, V. Leftwich, S. Weems, C. L. Kucera, L. Bouchard, and M. Game. "Strip mine reclamation: Production and decomposition of plant litter," in *Ecology and coal resource development, Vol. 2.* M. K. Wali, Ed. (New York: Pergamon Press, 1979), pp. 670–676.

Carrello, E. M. "Ten-year summary of environmental monitoring on coal mine spoil amended with sludge. The Status of Municipal Sludge Management for the 1990's," Proceedings of the Water Pollution Control Federation, Alexandria, VA (1990), 9–1 to 9–19.

Cavey, J. V. and J. A. Bowles. "Use of sewage sludge to improve taconite tailings as a medium for plant growth," in *Land Reclamation and Biomass Production with Municipal Wastewater and Sludge,* W. E. Sopper, E. M. Seaker, and R. K. Bastian, Eds. (University Park, PA: The Pennsylvania State University Press, 1982), pp. 400–409.

Chaney, R. "Crop and food chain effects of toxic elements in sludges and effluents," in *Recycling Municipal Sludges and Effluents on Land,* (Washington, DC: National Assoc. of State Univ. and Land Grant Colleges, 1973), pp. 129–141.

Chang, F. H. and F. E. Broadbent. "Influence of trace metals on some soil nitrogen transformations," *J. Environ. Qual.* 11:1–4 (1982).

Chapman, H. D. "Diagnostic criteria for plants and soils," University of California, Division of Agricultural Sciences, Riverside, CA (1982).

Church, D. C. and W. G. Pond. "Basic animal nutrition and feeding," (Portland, OR: Schultz/Wack/Weir, 1974).

Cocke, C. L. and K. W. Brown. "The effect of sewage on the physical properties of lignite overburden," *Reclam. Reveg. Res.* G:83–93 (1987).

Coker, E. G., R. D. Davis, J. E. Hall, and C. H. Carlton-Smith. "Field experiments on the use of consolidated sewage sludge for land reclamation: Effects on crop yield and composition and soil conditions," 1976-1981. Technical Report TR 183, Water Research Centre, Medmenham, U.K. (1982) p. 83.

Corey, R. B., L. D. King, C. Lue-Hing, D. S. Fanning, J. J. Street, and J. M. Walker. "Effects of sludge properties on accumulation of trace elements by crops," in *Land Application of Sludge,* A. L. Page, T. G. Logan, and J. A. Ryan, Eds. (Chelsea, MI: Lewis Publishers, Inc., 1987), pp. 25–51.

Council for Agricultural Science and Technology. "Application of sewage sludge to cropland: Appraisal of potential hazards of the heavy metals to plants and animals," Report No. 64. Office of Water Programs, U.S. EPA Report-430/9-76-013 (1976).

Council for Agricultural Science and Technology. "Effects of sewage sludge on the cadmium and zinc content of crops," CAST No. 83. Ames, IA (1980), p. 77.

Covey, J. N. "Fly ash disposal with sludge," Proc. Municipal and Industrial Sludge Utilization and Disposal, Washington, DC (1980).

Cundell, A. M. "The role of microorganisms in the revegetation of stripmined land in Western United States," *J. Range Manage.* 30:299–305 (1977).

Cunningham, J. D., D. R. Keeney, and J. A. Ryan. "Yield and metals composition of corn and rye grown on sewage-sludge-amended spoil," *J. Environ. Qual.* 4:448–454 (1975).

Curnow, R. D., W. A. Tolin, and D. W. Lynch. "Ecological and land-use relationships of toxic metals in Ohio's terrestrial vertebrate fauna," in *Biological Implications of Metals in the Environment,* H. Drucker and R. E. Wildung, Eds. (Richland, WA: Energy Research and Development Administration, 1977) pp. 578–594.

Davis, R. D. and P. H. T. Beckett. "Upper critical levels of toxic elements in plants II. Critical levels of copper in young barley, wheat, rape, lettuce, and ryegrass and of nickel and zinc in young barley and ryegrass," *New Phytol.* 80:23–32 (1978).

Daniels, W. L. and K. Haering. "The feasibility of large-scale sewage sludge/compost utilization on Central Appalachian surface mined lands," in *Proceedings of the 1990 Mining and Reclamation Conference and Exhibition, Vol. I,* Skousen et al., Eds. (Morgantown, WV: West Virginia University, 1990), pp. 165–170.

Department of Environment. "Survey of derelict and despoiled land in England," Planning and Regional Countryside Directorate, London, U.K. (1974).

Department of the Environment/National Water Council. "Report of the sub-committee on the disposal of sewage sludge to land," DOE Standing Technical Committee Report 20. London, U.K. (1981).

Doelman, P. and L. Haanstra. "Effect of lead on soil respiration and dehydrogenase activity," *Soil. Biol. Biochem.* 11:475–479 (1979).

Down, C. G. and J. Stocks. *Environmental Impact of Mining,* (London: Applied Sciences Publishers, 1977).

Dressler, R. L., G. L. Storm, W. M. Tzilkowski, and W. E. Sopper. "Heavy metals in cottontail rabbits on mined land treated with sewage sludge," *J. Environ. Qual.* 15:278–281 (1986).

Epstein, E., D. B. Keane, J. J. Meisinger, and J. O. Legg. "Mineralization of nitrogen from sewage sludge and sludge compost," *J. Environ. Qual.* 7:217–221 (1978).

Federal Register. "Criteria for Classification of Solid Waste Disposal Facilities and Practices: Final, Interim Final, and Proposed Regulations," 40 CFR Part 257, Federal Register 44 No. 179: 53438–69, Federal Register. September 13, 1979.

Federal Register. "Surface Coal Mining and Reclamation Permanent Program Regulations; Revegetation," Federal Register 47 No. 56: 12596-604. Part VI, Federal Register. March 23, 1982.

Federal Register. "Proposed Standards for the Disposal of Sewage Sludge," 40 CFR Parts 257 and 503, Federal Register 54 No. 23: 5746-5902, Federal Register. February 6, 1989.

Federal Register. "Standards for the Use and Disposal of Sewage Sludge, " Final Rule 40 CFR Part 503, Federal Register 58 No. 32, February 19, 1993.

Feuerbacher, T. A., R. I. Barnhisel, and M. D. Ellis. "Utilization of sewage sludge as a spoil amendment in the reclamation of lands surface mined for coal," in Proc. Symp. on Surface Mining Hydrol., Sedimentol., and Reclamation, University of Kentucky, Lexington (1980), pp. 187–192.

Fitzgerald, P. R. "Recovery and utilization of strip-mined land by application of anaerobically digested sludge and livestock grazing," in *Utilization of Municipal Sewage Effluent and Sludge on Forest and Disturbed Land,* W. E. Sopper and S. N. Kerr, Eds. (University Park, PA: The Pennsylvania State University Press, 1979), pp. 497–506.

Fitzgerald, P. R. "Effects of natural exposure of cattle and swine to anaerobically digested sludge," in *Land Reclamation and Biomass Production with Municipal Wastewater and Sludge,* W. E. Sopper, E. M. Seaker, and R. K. Bastian, Eds. (University Park, PA: The Pennsylvania State University Press, 1982), pp. 359–67.

Follett, R. H., L. S. Murphey, and R. L. Donahue. *Fertilizers and Soil Amendment,* (Englewood Cliffs, NJ: Prentice-Hall, Inc., 1981), p. 557.

Frank, A. "Mining waste, mine reclamation, and municipal sludge," *Sludge Magazine,* July-Aug. 1978, 23–27.

Franks, W. A., M. Persinger, A. Iob, and P. Inyangetor. "Utilization of sewage effluent and sludge to reclaim soil contaminated by toxic fumes from a zinc smelter," in *Land Reclamation and Biomass Production with Municipal Wastewater and Sludge,* W. E. Sopper, E. M. Seaker, and R. K. Bastian, Eds. (University Park, PA: The Pennsylvania State University Press, 1982), pp. 219–51.

Fresquez, P. R. and E. F. Aldon. "Microbial reestablishment and the diversity of fungal genera in reclaimed coal mine spoils and soils," *Reclam. Reveg. Res.* 4:245–258 (1986).

Fresquez, P. R. and W. C. Lindemann. "Soil and rhizosphere microorganisms in amended coal mine spoils," *Soil Sci. Soc. Am. J.* 46:751–755 (1982).

Fresquez, P. R., R. E. Francis, and G. L. Dennis. "Sewage sludge effects on soil and plant quality in a degraded, semiarid grassland," *J. Environ. Qual.* 19(2):324–329 (1990a).

Fresquez, P. R., R. E. Francis, and G. L. Dennis. "Soil and vegetation responses to sewage sludge on a degraded semiarid broom snakeweed/blue gamma plant community," *J. Range Manage.* 43(4):325–331 (1990b).

Gadd, G. M. and A. J. Griffiths. "Microorganisms and heavy metal toxicity," *Microb. Ecol.* 4:303–317 (1978).

Gaffney, G. R. and R. Ellerston. "Ion uptake of redwinged blackbirds nesting on sludge-treated spoils," in *Utilization of Municipal Sewage Effluent and Sludge on Forest and Disturbed Land,* W. E. Sopper and S. N. Kerr, Eds. (University Park, PA: The Pennsylvania State University Press, 1979), pp. 407–516.

Gaskin, D. A., W. Hannel, A. J. Palazzo, R. E. Bates, and L. E. Stanley. "Utilization of sewage sludge for terrain stabilization in cold regions," Special Report 77–37, U.S. Army Cold Regions Research and Engineering Laboratory, Hanover, NH (1977).

Griebel, G. E., W. H. Armiger, J. F. Parr, D. W. Steck, and J. A. Adam. "Use of composted sewage sludge in revegetation of surface-mined areas," in *Utilization of Municipal Sewage Effluent and Sludge on Forest and Disturbed Land,* W. E. Sopper and S. N. Kerr, Eds., (University Park, PA: The Pennsylvania State University Press, 1979), pp. 293–306.

Haering, K. and W. Lee Daniels. "Development of new technologies for the utilization of municipal sewage sludge on surface mined lands," Powell River Project Symposium and Progress Reports, Clinch Valley College, Wise, VA (1991), pp. 68–74.

Haghiri, F. and P. Sutton. "Vegetation establishment on acidic mine spoils as influenced by sludge application," in *Land Reclamation and Biomass Production with Municipal Wastewater and Sludge,* W. E. Sopper, E. M. Seaker, and R. K. Bastian, Eds. (University Park, PA: The Pennsylvania State University Press, 1982), pp. 433–46.

Hall, J. E., Ed. *Alternative Uses For Sewage Sludge,* (Oxford, U.K.: Pergamon Press, 1991), p. 387.

Hartenstein, R. E., E. F. Neuhausen, and A. Narahara. "Effects of heavy metal and other elemental additives to activated sludge on growth of *Eisenia foetida*," *J. Environ. Qual.* 10:372–367 (1981).

Haynes, R. J. and W. D. Klimstra. "Illinois lands surface mined for coal," Cooperative Wildlife Res. Lab., Southern Illinois University, Carbondale, IL (1975).

Hill, R. D., K. Hinkle, and R. S. Klingensmith. "Reclamation of orphaned mined lands with municipal sludges — case studies," in *Utilization of Municipal Sewage Effluent and Sludge on Forest and Disturbed Land,* W. E. Sopper and S. N. Kerr, Eds. (University Park, PA: The Pennsylvania State University Press, 1979), pp. 423–43.

Hinesly, T. D. and K. E. Redborg. "Long-term use of sewage sludge on agricultural and disturbed lands," Municipal Environmental Research Lab, Cinncinnati, OH, U.S. EPA Report-600/2-84-126 (1984), p. 78.

Hinesly, T. D., D. E. Redborg, E. L. Ziegler, and I. H. Rose-Innes. "Effects of chemical and physical changes in strip-mined spoil amended with sewage sludge on the uptake of metals by plants," in *Land Reclamation and Biomass Production with Municipal Wastewater and Sludge,* W. E. Sopper, E. M. Seaker, and R. K. Bastian, Eds. (University Park, PA: The Pennsylvania State University Press, 1982), pp. 339–352.

Hinesly, T. D., E. L. Ziegler, and G. L. Barrett. "Residual effects of irrigated corn with digested sewage sludge," *J. Environ. Qual.* 8:35–38 (1979a).

Hinesly, T. D., L. G. Hansen, E. L. Ziegler, and G. L. Barrett. "Effects of feeding corn grain produced on sludge-amended soil to pheasants and swine," in *Utilization of Municipal Sewage Effluent and Sludge on Forest and Disturbed Land,* W. E. Sopper and S. N. Kerr, Eds. (University Park, PA: The Pennsylvania State University Press, 1979b), pp. 483–496.

Hinesly, T. D., K. E. Redborg, R. I. Pietz, and E. L. Ziegler. "Cadmium and zinc uptake by corn (*Zea mays* L.) with repeated applications of sewage sludge," *J. Agric. and Food Chem.* 32:155–163 (1984).

Hinkle, K. R. "Use of Municipal sludge in the reclamation of abandoned pyrite mines in Virginia," in *Land Reclamation and Biomass Production with Municipal Wastewater and Sludge,* W. E. Sopper, E. M. Seaker, and R. K. Bastian, Eds. (University Park, PA: The Pennsylvania State University Press, 1982), pp. 421–32.

Hornby, W. J., K. W. Brown, and J. C. Thomas. "Nitrogen mineralization potentials of revegetated lignite overburden in the Texas gulf coast," *Soil Sci. Soc. Amer. Proc.* 50:1484–1489 (1986).

Hornick, S. B. "Crop production on waste amended gravel spoils," in *Land Reclamation and Biomass Production with Municipal Wastewater and Sludge,* W. E. Sopper, E. M. Seaker, and R. K. Bastian, Eds. (University Park, PA: The Pennsylvania State University Press, 1982), pp. 207–218.

Jenny, H. *The Soil Resource* (New York, NY: Springer-Verlag, 1980).

John, M. K., C. J. Vanlaerhoven, and H. H. Chuah. "Factors affecting plant uptake and phytotoxicity of cadmium added to soils," *J. Environ. Sci. Technol.* 6(12):1005–1009 (1972).

Johnson, M. S., R. D. Roberts, M. Hutton, and M. J. Inskip. "Distribution of lead, zinc and environments," *Oikos* 30:153–159 (1978).

Jones, J. B. "Plant tissue analysis for micronutrients," in *Micronutrients in Agriculture,* J. J. Mortvedt, P. M. Giordiano, and W. L. Lindsay, Eds. (Madison, WI: Soil Sci. Soc. Am., 1972), pp. 319–346.

Jones, M. and R. S. Cunningham. "Sludge used for strip mine restoration at Palzo: Project development and compliance water quality monitoring," in *Utilization of Municipal Sewage Effluent and Sludge on Forest and Disturbed Land,* W. E. Sopper and S. N. Kerr, Eds. (University Park, PA: The Pennsylvania State University Press, 1979), pp. 269–278.

Joost, R. E., J. H. Jones, and F. J. Olsen. "Physical and chemical properties of coal refuse as affected by deep incorporation of sewage sludge and/or limestone," in Proc. Symp. on Surface Mining Hydrol., Sedimentol., and Reclamation, University of Kentucky, Lexington (1981), pp. 307–312.

Joost, R. E., F. J. Olsen, and J. H. Jones. "Revegetation and minespoil development of coal refuse amended with sewage sludge and limestone," J. Environ. Qual. 16:65–68 (1987).

Jurgensen, M. F. "Microorganisms and the reclamation of mine wastes," in Forest Soils and Land Use, C. T. Youngbert, Ed. (Fort Collins, CO: Colorado State University, 1978), pp. 251–286.

Kardos, L. T., W. E. Sopper, B. R. Edgerton, and L. E. DiLissio. "Sewage effluent and liquid digested sludge as aids to revegetation of strip mine spoil and anthracite coal refuse banks," in Utilization of Municipal Sewage Effluent and Sludge on Forest and Disturbed Land, W. E. Sopper and S. N. Kerr, Eds. (University Park, PA: The Pennsylvania State University Press, 1979), pp. 315–32.

Keefer, R. F., R. N. Singh, O. L. Bennett, and D. J. Horvath. "Chemical composition of plants and soils from revegetated mine soils," Proc. Symposium on Surface Mining, Hydrology, Sedimentology, and Reclamation, University of Kentucky, Lexington, KY (1983), pp. 155–161.

Kerr, S. N., W. E. Sopper, and B. R. Edgerton. "Reclaiming anthracite refuse banks with heat-dried sewage sludge," in Utilization of Municipal Sewage Effluent and Sludge on Forest and Disturbed Land, W. E. Sopper and S. N. Kerr, Eds. (University Park, PA: The Pennsylvania State University Press, 1979), pp. 333–352.

Kormanik, P. P. and R. C. Schultz. "Significance of sewage sludge amendments to borrow pit reclamation with sweetgum and fescue," U.S. For. Serv., Res. Note SE-329, (1985), p. 7.

Knezek, B. D. and R. H. Miller, Eds. Application of sludges and wastewaters on agricultural land: A planning and educational guide, North Central Regional Research Publication 235 (Wooster: Ohio Agricultural Research and Development Center, 1976). (Reprinted by the U.S. Environmental Protection Agency as MCD-35).

Lawrey, J. C. "Soil fungal population and soil respiration in habitats variously influenced by coal strip-mining," Environ. Pollut. 14:195–205 (1977).

Leisman, G. A. "A vegetation and soil chronosequence on the Mesabi Iron Range spoil banks," Minnesota. Ecol. Monogr. 27:221–245 (1957).

Lejcher, T. R. and S. H. Kunkle. "Restoration of acid spoil banks with treated sewage sludge, in Recycling Treated Municipal Wastewater and Sludge through Forest and Cropland, W. E. Sopper and L. T. Kardos, Eds. (University Park, PA: The Pennsylvania State University Press, 1973), pp. 184–199.

Lighthart, B., J. Baham, and V. V. Volk. "Microbial respiration and chemical speciation in metal-amended soils," J. Environ. Qual. 12:543–548 (1983).

Logan, T. J. and R. L. Chaney. "Utilization of municipal wastewater and sludge on land–metals," in Utilization of Municipal Wastewater and Sludge on Land, A. L. Page et al., Eds. (Riverside, CA: University of California, 1983), pp. 235–326.

Lue-Hing, C., S. J. Sedita, and K. C. Rao. "Viral and bacterial levels resulting from the land application of digested sludge," in Utilization of Municipal Sewage Effluent and Sludge on Forest and Disturbed Land, W. E. Sopper and S. N. Kerr, Eds. (University Park, PA: The Pennsylvania State University Press, 1979), pp. 445–462.

Magdoff, F. R. and F. W. Chromec. "N mineralization from sewage sludge," J. Environ. Sci. Health A12:191–201 (1977).

Martin, J. P., J. O. Ervin, and R. A Shepherd. "Decomposition of the iron, aluminum, zinc, and copper salts for complexes of some microbial and plant polysaccharides in soil," Soil Sci. Soc. Am. Proc. 30:196–200 (1966).

Martin, W. E. and J. E. Matocha. "Plant analysis as an aid in the fertilization of forage crops," in *Soil testing and plant analysis,* L. A. Walsh and J. D. Beaton, Eds. (Madison, WI: Soil Sci. Soc. Am., 1973), pp. 393–426.

Mathias, E. L., O. L. Bennett, and P. E. Lundberg. "Use of sewage sludge to establish tall fescue on strip mine spoils in West Virginia," in *Utilization of Municipal Sewage Effluent and Sludge on Forest and Disturbed Land,* W. E. Sopper and S. N. Kerr, Eds. (University Park, PA: The Pennsylvania State University Press, 1979), pp. 307–314.

Mathur, S. P., H. A. Hamilton, and M. P. Levesque. "The mitigating effect of residual fertilizer copper on the decomposition of an organic soil in situ," *Soil Sci. Soc. Am. J.* 43:200–203 (1979).

McBride, F. D., C. Chavengsaksengkram, and D. H. Urie. "Sludge-treated coal mine spoils increase heavy metals in cover crops," Research Notes NC-221, St. Paul, MN: North Central Forest Experimental Station (1977).

McCormick, L. H. and F. Y. Borden. "Percolate from spoils treated with sewage effluent and sludge," in *Ecology and Reclamation of Devastated Land, Vol. I,* R. J. Hutnick and G. Davis, Eds. (New York: Gordon and Breach, 1973), pp. 239–50.

Melsted, S. W. "Soil-plant relationships," in *Recycling Municipal Sludges and Effluents on Land,* (Washington, DC: National Assoc. of State Univ. and Land-Grant Colleges, 1973), pp. 121–128.

Metcalfe, B. "The use of consolidated sewage sludge as a soil substitute in colliery spoil reclamation," *Water Poll. Control.* 83:288–299 (1984).

Metcalfe, B. and J. C. Lavin. "Consolidated sewage sludge as a soil substitute in colliery spoil reclamation," in *Alternative Uses for Sewage Sludge,* J. E. Hall, Ed. (Oxford, U.K.: Pergamon Press 1991), pp. 83–96.

Miller, R. H. "Soil microbiology aspects of recycling sewage sludges and waste effluents on land," in *Recycling municipal sludges and effluents on land,* (Washington, DC: National Assoc. of State Universities and Land-Grant Colleges, 1974), pp. 79–90.

Miller, R. M. and R. E. Cameron. "Microbial ecology studies at two coal mine refuse sites in Illinois," Argonne National Laboratory Rep. ANL/LRP-3. Natl. Technical Information Serv., U.S. Department of Commerce, Springfield, VA (1978).

Miller, R. M. and S. W. May. "Staunton 1 reclamation demonstration project, Progress report II," Argonne National Laboratory Rep. ANL/LRP-4. Natl. Technical Information Serv., U.S. Department of Commerce, Springfield, VA (1979).

Mills, A. L. "Acid mine waste drainage: Microbial impact on the recovery of soil and water ecosystems," in *Soil Reclamation Processes,* R. L. Tate and D. A. Klein, Eds. (New York: Marcel Dekker, 1985), pp. 35–81.

Morin, M. "Heavy metal concentrations in three year old trees grown on sludge-amended surface mine spoil, in Proc. Symp. on Surface Mining Hydrol., Sedimentol., and Reclamation, University of Kentucky, Lexington (1981), pp. 297–306.

Morrison, D. G. and J. Hardell. "The response of native herbaceous prairie species on iron-ore tailings under different rates of fertilizer and sludge application," in *Land Reclamation and Biomass Production with Municipal Wastewater and Sludge,* W. E. Sopper, E. M. Seaker and R. K. Bastian, Eds. (University Park, PA: The Pennsylvania State University Press, 1982), pp. 410–420.

Murray, D. T., S. A. Townsend, and W. E. Sopper. "Using sludge to reclaim mine land," *Biocycle* 22(3):48–55 (1981).

National Research Council. *Mineral Tolerance of Domestic Animals* (Washington, DC: National Academy of Science, 1980).

Office of Surface Mining Reclamation and Enforcement. "Surface Coal Mining Reclamation: 10 years of Progress, 1977–1987," U.S. Department of Interior, Washington, DC (1987), p. 48.

Ollier, C. *Weathering* (Edinburgh, Scotland: Oliver and Boyd, 1969).

Palazzo, A. J. "Reclamation of acidic dredge spoils with sewage sludge and lime at the Chesapeake and Delaware canal," Special Report 77–19, U.S. Army Cold Regions Research and Engineering Laboratory, Hanover, NH (1977).

Palazzo, A. J. and R. W. Duell. "Responses of grasses and legumes to soil pH," *Agron. J.* 66:678–682 (1974).

Palazzo, A. J. and C. M. Reynolds. "Long-term changes in soil and plant metal concentrations in an acidic dredge disposal site receiving sewage sludge," *Water Air Soil Poll.* 57-58:839-848 (1991).

Palazzo, A. J., S. D. Rindge, and D. A. Gaskin. "Revegetation at two construction sites in New Hampshire and Alaska," CRREL Report 80-3, U.S. Army Cold Regions Research and Engineering Laboratory, Hanover, NH (1980).

Paone, J., P. Struthers, and W. Johnson. "Extent of disturbed lands and major reclamation problems in the United States, in *Reclamation of Drastically Disturbed Lands,* F. W. Schaller and P. Sutton, Eds. (Madison, WI: American Society of Agronomy, 1978).

Parkinson, D., S. Visser, R. M. Danielson, and J. Zak. "Restoration of fungal activity in tailing sand (oil sands). in *Soil biology as related to land use practices,* D. L. Dindal, Ed., Proc. 7th Int. Soil Zool. Colloq. Int. Soc. Soil Sci., Syracuse , NY. 29 July–30 Aug. 1979. U.S. EPA, Washington, DC (1980).

Pennsylvania Crop Reporting Service. "Crop and livestock annual summary 1980," Pennsylvania Department of Agriculture, Harrisburg, PA (1980).

Pennsylvania Department of Environmental Resources. "Interim guidelines for sewage sludge use for land reclamation," in The Rules and Regulations of the Department of the Environmental Resources, Commonwealth of Pennsylvania, Chap. 75, Subchap. C., Sec. 75.32 (1977).

Pennsylvania Department of Environmental Resources. Guidelines for sewage sludge land reclamation, in Rules and Regulations of the Department of Environmental Resources, Commonwealth of Pennsylvania, Chapter 275 (1988).

"Dairy Reference Manual," Pennsylvania State University, College of Agriculture, University Park, PA (1970).

Peterson, J. R., C. Lue-Hing, J. Gschwind, R. I. Pietz, and D. R. Zenz. "Metropolitan Chicago's Fulton County sludge utilization program," in *Land Reclamation and Biomass Production with Municipal Wastewater and Sludge,* W. E. Sopper, E. M. Seaker, and R. K. Bastian, Eds., (University Park, PA: The Pennsylvania State University Press, 1982), pp. 322–338.

Peterson, J. R., R. I. Pietz, and C. Lue-Hing. "Water, soil, and crop quality of Illinois coal mine spoils amended with sewage sludge," in *Utilization of Municipal Sewage Effluent and Sludge on Forest and Disturbed Land,* W. E. Sopper and S. N. Kerr, Eds. (University Park, PA: The Pennsylvania State University Press, 1979), pp. 359–368.

Pietz, R. I., C. R. Carlson, Jr., J. R. Peterson, D. R. Zenz, and C. Lue-Hing. "Application of sewage sludge and other amendments to coal refuse material: I. Effects on chemical composition," *J. Environ. Qual.* 18:164–169 (1989a).

Pietz, R. I., C. R. Carlson, Jr., J. R. Peterson, D. R. Zenz, and C. Lue-Hing. "Application of sewage sludge and other amendments to coal refuse material: II. Effects on vegetation," *J. Environ. Qual.* 18:169–173 (1989b).

Pietz, R. I., C. R. Carlson, Jr., J. R. Peterson, D. R. Zenz, and C. Lue-Hing. "Application of sewage sludge and other amendments to coal refuse material: III. Effects on percolate water composition," *J. Environ. Qual.* 18:174–179 (1989c).

Pietz, R. I., J. R. Peterson, J. E. Prater, and D. R. Zenz. "Metal concentrations in earthworms from sewage sludge-amended soils at a strip mine reclamation site," *J. Environ. Qual.* 13:651–654 (1984).

Powell, J. L., R. I. Barnhisel, F. A. Craig, J. R. Armstrong, M. L. Ellis, and W. O. Thom. "The use of organic amendments in the restoration of prime farmland," in *New Horizons in Mined Land Reclamation*, J. Harper and B. Plass Eds., Proc. Natl. Meet. Am. Soc. Surface Mining and Reclamation, Princeton, WV (1988), pp. 55–60.

Premi, P. R. and A. H. Cornfield. "Effects of additives of copper, manganese, zinc, and chromium compounds on ammonification and nitrification during incubation of soil," *Plant Soil* 31:345–352 (1969).

Pulford, I. D. "Sewage sludge as an amendment for reclaimed colliery spoil," in *Alternative Uses for Sewage Sludge*, J. E. Hall, Ed. (Oxford, U.K.: Pergamon Press, 1991), pp. 41–54.

Pulford, I. D., T. H. Flowers, S. H. Shah, and T. A. B. Walker. "Supply and turnover of N, P, and K in reclaimed coal mine waste in Scotland. Mine Drainage and Surface Mine Reclamation, Vol. II," U.S.D.I. Bureau of Mines Information Circular 9184, (1988) pp. 228–235.

Reeder, J. D. and W. A. Berg. "Nitrogen mineralization and nitrification in a cretaceous shale and coal mine spoil," *Soil Sci. Soc. Am. Proc.* 41:922–927 (1977).

Roberts, J. A., W. L. Daniels, J. C. Bell, and D. C. Martens. "Tall fescue production and nutrient status on southwest Virginia mine soils," *J. Environ. Qual.* 17:55–62 (1988).

Roth, P. L., B. D. Jayko, and G. T. Weaver. "Initial survival and performance of woody plant species on sludge-treated spoils of the Palzo Mine," in *Utilization of Municipal Sewage Effluent and Sludge on Forest and Disturbed Land*, W. E. Sopper and S. N. Kerr, Eds. (University Park, PA: The Pennsylvania State University Press, 1979), pp. 389–394.

Roth, P. L., G. T. Weaver, and M. Morin. "Restoration of woody ecosystem on a sludge-amended devastated mine-site," in *Land Reclamation and Biomass Production with Municipal Wastewater and Sludge*, W. E. Sopper, E. M. Seaker, and R. K. Bastian, Eds. (University Park, PA: The Pennsylvania State University Press, 1982), pp. 368–385.

Sabey, B. R., N. N. Agbim, and D. C. Markstrom. "Land application of sewage sludge: IV. Wheat growth, N content, N fertilizer value, and N use efficiency as influenced by sewage sludge and wood waste mixtures," *J. Environ. Qual.* 6:52–58 (1977).

Sabey, B. R., R. L. Pendleton, and B. L. Webb. "Effect of municipal sewage sludge application on growth of two reclamation shrub species in Copper mine spoils," *J. Environ. Qual.* 19:580–586 (1990).

Scanlon, D. H., C. Duggan, and S. D. Bean. "Evaluation of municipal compost for strip mine reclamation," *Compost Sci.* 14:3 (1973).

Schafer, W. M., G. A. Neilson, and W. D. Nettleton. "Minespoil genesis morphology in a spoil chronosequence in Montana," *Soil Sci. Soc. Am. J.* 44:802–807 (1980).

Schaller, F. W. and P. Sutton, Eds. *"Reclamation of Drastically Disturbed Lands,"* ASA-CSSA-SSA, Madison, WI (1978) p. 742.

Schneider, K. R., R. J. Wittwer, and S. B. Carpenter. "Trees respond to sewage sludge in reforestation of acid spoil," Proc. Symp. on Surface Mining Hydrol., Sedimentol., and Reclamation, University of Kentucky, Lexington (1981), pp. 291–296.

Seaker, E. M. "Zinc, copper, cadmium and lead in mine spoil, water, and plants from reclaimed mine land with sewage sludge," *Water Air Soil Poll.* 57-58:849–859 (1991).

Seaker, E. M. and W. E. Sopper. "Production and quality of forage vegetation grown on municipal sludge-amended mine spoil," Proc. Symp. Surface Mining Hydrol., Sedimentol., and Reclamation, University of Kentucky, Lexington (1982), pp. 715–720.

Seaker, E. M. and W. E. Sopper. "Reclamation of deep mine refuse banks with municipal sewage sludge," *Waste Manage. Res.*, Copenhagen, Denmark, 1:309–322 (1983).

Seaker, E. M. and W. E. Sopper. "Reclamation of bituminous strip mine spoil banks with municipal sewage sludge," *Reclam. Reveg. Res.* 3:87–100 (1984).

Seaker, E. M. and W. E. Sopper. "Municipal sludge for minespoil reclamation: I. Effects on microbial populations and activity," *J. Environ. Qual.* 17:591–597 (1988a).

Seaker, E. M. and W. E. Sopper. "Municipal sludge for minespoil reclamation: II. Effects on organic matter," *J. Environ. Qual.* 17:598–602 (1988b).

Sedita, S. J., P. O'Brien, J. J. Bertucci, C. Lue-Hing, and D. R. Zenz. "Public health aspects of digested sludge utilization," in *Land as a Waste Management Alternative,* Proc. 1976 Cornell, Agric. Waste Mgmt. Conf., R. C. Loehr, Ed. (Ann Arbor, MI: Ann Arbor Science Publishers, 1977).

Segal, W. and R. L. Mancinelli. "Extent of regeneration of the microbial community in reclaimed spent oil shale land," *J. Environ. Qual.* 16:44–48 (1987).

Shackette, H. and J. Boergen. "Elemental concentrations in soils and other surficial materials of the conterminous United States," U.S. Geological Survey Professional Paper 1270, U.S. Geological Survey, Alexandria, VA (1984).

Sharma, R. P. "Soil-plant-animal distribution of cadmium in the environment," in *Cadmium in the environment, Part 1. Ecological Cycling,* J. O. Nriagu, Ed. (New York, NY: John Wiley and Sons, Inc. 1981), pp. 587–605.

Skousen, J. G. "Effects of sewage sludge application on a revegetated minesoil in West Virginia. Mine drainage and surface mine reclamation, Vol. II," U.S.D.I. Bureau of Mines Information Circular 9184 (1988) pp. 214–220.

Smith, R. M., E. H. Tryon, and E. H. Tyner. "Soil development on mine spoil," West Virginia Univ. Exp. Stn. Bull. 604T, Morgantown, WV (1971).

Sommers, L. E., D. W. Nelson, and K. J. Yost. "Variable nature of chemical composition of sewage sludges," *J. Environ. Qual.* 5:303–306 (1976).

Sommers, L. E., D. W. Nelson, A. W. Kirleis, S. D. Strachan, J. C. Inman, S. A. Boyd, J. G. Graveel, and A. D. Behel. "Characterization of sewage sludge and sewage sludge-soil systems," U.S.EPA Report-600/2-84-046 (1984).

Sommers, L., V. Van Volk, P. M. Giordano, W. E. Sopper, and R. Bastian. "Effects of soil properties on accumulation of trace elements in crops," in *Land Application of Sludge,* A. L. Page, T. G. Logan, and J. A. Ryan, Eds., (Chelsea, MI: Lewis Publishers, Inc., 1987), pp. 5–24.

Sopper, W. E. Strip mine reclamation project — Kennell I site, Institute for Research on Land and Water Resources, The Pennsylvania State University, University Park, PA (1982a), p. 17.

Sopper, W. E. "Strip mine reclamation project — Mast site," Institute for Research on Land and Water Resources, The Pennsylvania State University, University Park, PA (1982b), p. 24.

Sopper, W. E. "Strip mine reclamation project — Leasure site," Institute for Research on Land and Water Resources, The Pennsylvania State University, University Park, PA (1982c), p. 15.

Sopper, W. E. "Strip mine reclamation project — Beachy I site," Institute for Research on Land and Water Resources, The Pennsylvania State University, University Park, PA (1982d), p. 18.

Sopper, W. E. "Strip Mine Reclamation Project — Black site," Institute for Research on Land and Water Resources, The Pennsylvania State University, University Park, PA (1982e), p. 15.

Sopper, W. E. "Reclamation of the Palmerton superfund site," Proceedings of the 1987 National Symposium on Mining, Hydrology, Sedimentology and Reclamation, University of Kentucky, Lexington, KY (1987), pp. 171–178.

Sopper, W. E. "Revegetation of a contaminated zinc smelter site," *Lands. Urban Plan.* 17:241–250 (1989a).

Sopper, W. E. "Reforestation of the Palmerton zinc smelter superfund site," Healthy Forests, Healthy World, Proceedings of the 1988 Society of American Foresters National Convention, Bethesda, MD (1989b), pp. 205–209.

Sopper, W. E. "Bio-remediation of the Palmerton superfund site," Proceedings of 6th National Materials Conference on Hazardous Wastes and Hazardous Materials, Hazardous Materials Control Research Institute, Silver Springs, MD (1989d), pp. 322–326.

Sopper W. E. "Revegetation of burned anthracite coal refuse banks using municipal sludge," Proceedings of the 1990 National Symposium on Mining, University of Kentucky, Lexington, KY (1990a), pp. 37–42.

Sopper, W.E. "Use of sewage sludge-fly ash mixtures to revegetate steep coal waste banks," Environmental Contamination, CEP Consultants Ltd., Edinburgh, U.K. (1990b), pp. 51–57.

Sopper, W. E. "Long-term effects of reclamation of deep coal mine refuse banks with municipal sewage sludge," Proceedings of the 1991 National Symposium on Mining, University of Kentucky, KY (1991a), pp. 55–62.

Sopper, W. E. "Revegetation of steep coal waste banks using a sewage sludge-fly ash amendment," Proceedings of the 1991 National Meeting of the American Society of Surface Mining and Reclamation, Vol. II, W. R. Oaks and J. Bowden, Eds., Princeton, WV (1991b), pp. 413–420.

Sopper, W. E. and S. N. Kerr, Eds. *Utilization of Municipal Sewage Effluent and Sludge on Forest and Disturbed Land,* (University Park, PA: The Pennsylvania State University Press, 1979), p. 537.

Sopper, W. E. and S. N. Kerr. "Strip mine reclamation demonstration project — Decker site," Institute for Research on Land and Water Resources, The Pennsylvania State University, University Park, PA (1980a), p. 18.

Sopper, W. E. and S. N. Kerr. "Strip mine reclamation demonstration project — Soberdash site," Institute for Research on Land and Water Resources. The Pennsylvania State University, University Park, PA (1980b), p. 22.

Sopper, W. E. and S. N. Kerr. "Revegetating strip-mined land with municipal sewage sludge," Project Summary U.S.EPA Report-600/52-81-182 (1981), p. 7.

Sopper, W. E. and S. N. Kerr. "Mine land reclamation with municipal sludge - Pennsylvania's demonstration program," in *Land Reclamation and Biomass Production with Municipal Wastewater and Sludge,* W. E. Sopper, E. M. Seaker, and R. K. Bastian, Eds. (University Park, PA: The Pennsylvania State University Press, 1982), pp. 55–74.

Sopper, W. E. and J. M. McMahon. "Greening of Palmerton Blue Mountain," *Biocycle* 28(4):47–51 (1987).

Sopper, W. E. and J. M. McMahon, "Revegetation of a superfund site, Part I," *Biocycle* 29(7):57–60 (1988a).

Sopper, W. E. and J. M. McMahon. "Revegetation of a superfund site, Part II," *Biocycle* 29(8):64–66 (1988b).

Sopper, W. E., S. N. Kerr, E. M. Seaker, W. F. Pounds, and D. T. Murray. "The Pennsylvania program for using municipal sludge for mine land reclamation," Proc. Symp. Surface Mining Hydrol., Sedimentol., and Reclamation, University of Kentucky, Lexington, KY (1981), pp. 283–290.

Sopper, W. E. and E. M. Seaker. "Strip mine reclamation with municipal sludge," Final Technical Report on Contract No. CR807408010, Municipal Environmental Research Lab, Environmental Protection Agency, Cincinnati, OH (1982), p. 194.

Sopper, W. E. and E. M. Seaker. "A Guide for revegetation of mined land in eastern United States using municipal sludge," Institute for Research on Land and Water Resources, The Pennsylvania State University, University Park, PA (1983), p. 93.

Sopper, W. E. and E. M. Seaker. "Use of municipal sludge to reclaim mined land," *CRC Crit. Rev. Environ. Control* 13:227–271 (1984a).

Sopper, W. E. and E. M. Seaker. "Strip mine reclamation with municipal sludge," Project Summary, U.S.EPA Report-600/S2-84-035 (1984b), p. 6.

Sopper, W. E. and E. M. Seaker. "Sludge brings life to microbial communities," *Biocycle* 28(4):40–47 (1987).

Sopper, W. E. and E. M. Seaker. "Development of microbial communities on sludge-amended mine land," in *Innovative Approaches to Mine Land Reclamation,* C. L. Carlson and J. W. Swisher, Eds. (Carbondale, IL: Southern Illinois University Press, 1987b), pp. 659–680.

Sopper, W. E. and E. M. Seaker. "Rejuvenation of microbial communities on abandoned mine land amended with municipal sludge," in Proc. Symp. on Mining, Hydrology, Sedimentology and Reclamation, University of Kentucky, Lexington, KY (1988), pp. 199–206.

Sopper, W. E. and E. M. Seaker. "Long-term effects of a single application of municipal sludge on abandoned mine land," in Proc. of the 1990 Mining and Reclamation Conf. and Exhibition, Vol. II, West Virginia University, Morgantown, WV (1990), pp. 579–587.

Sopper, W. E., E. M. Seaker, and R. K. Bastian, Eds. *Land Reclamation and Biomass Production with Municipal Wastewater and Sludge,* (University Park, PA: The Pennsylvania State University Press, 1982), p. 524.

Stevenson, H. F. *Humus Chemistry,* (New York: John Wiley & Sons, 1982).

Stowasser, W. F. "Phosphate rock," in *Mineral facts and problems,* Bulletin 667, U.S.D.I., Bureau of Mines, Washington DC (1975).

Stroo, H. F. and E. M. Jencks. "Enzyme activity and respiration in mine spoils," *Soil Sci. Soc. Am. J.* 46:548–553 (1982).

Stucky, D. J. and J. H. Bauer. "Establishment, yield, and ion accumulation of several forage species on sludge-treated spoils of the Palzo Mine," in *Utilization of Municipal Sewage Effluent and Sludge on Forest and Disturbed Land,* W. E. Sopper and S. N. Kerr, Eds. (University Park, PA: The Pennsylvania State University Press, 1979), pp. 379–388.

Stucky, D. J. and T. S. Newman. "Effect of dried anaerobically digested sewage sludge on yield and elemental accumulation in tall fescue and alfalfa," *J. Environ. Qual.* 6(3):271–274 (1977).

Stucky, D. J., J. H. Bauer, and T. C. Lindsey. "Restoration of acidic mine spoils with sewage sludge," *Reclamation Rev.* 3:129–139 (1980).

Sundberg, W. J., D. L. Borders, and G. L. Albright. "Changes in soil microfungal populations in the Palzo strip mine spoil following sludge application," in *Utilization of Municipal Sewage Effluent and Sludge on Forest and Disturbed Land,* W. E. Sopper and S. N. Kerr, Eds. (University Park, PA: The Pennsylvania State University Press, 1979), pp. 463–470.

Sutton, P. Ohio Agricultural Research and Development Center, Wooster, OH, Personal communication (1980).

Sutton, P. and J. P. Vimmerstedt. "Treat stripmine spoils with sewage sludge," *Compost Sci.* 15(1):22–23 (1974).

Svoboda, D., G. Smout, G. T. Weaver, and P. L. Roth. "Accumulation of heavy metals in selected woody plant species on sludge-treated strip mine spoils at the Palzo Site, Shawnee National Forest," in *Utilization of Municipal Sewage Effluent and Sludge on Forest and Disturbed Land,* W. E. Sopper and S. N. Kerr, Eds. (University Park, PA: The Pennsylvania State University, 1979), pp. 395–406.

Tate, R. L., III. "Microorganisms, ecosystem disturbance and soil formation processes," in *Soil reclamation processes,* R. L. Tate, III and D. A. Klein, Eds. (New York: Marcel Dekker, 1985), pp. 1–33.

Terry, R. E., D. W. Nelson, and L. E. Sommers. "Nitrogen transformations in sewage sludge-amended soils as affected by soil environmental factors," *Soil Sci. Soc. Am. J.* 45:506–513 (1981).

Tomlinson, T. G. "Inhibition of nitrification in the activated sludge process of sewage disposal," *J. Appl. Bacteriol.* 29:266 (1966).

Topper, K. F. and B. R. Sabey. "Sewage sludge as a coal mine spoil amendment for revegetation in Colorado," *J. Environ. Qual.* 15:44–49 (1986).

Tunison, K. W., B. C. Bearce, and H. A. Menser, Jr. "The utilization of sewage sludge: bark screenings compost for the culture of blueberries on acid minespoil," in *Land Reclamation and Biomass Production with Municipal Wastewater and Sludge,* W. E. Sopper, E. M. Seaker, and R. K. Bastian, Eds. (University Park, PA: The Pennsylvania State University Press, 1982), pp. 195–206.

U.S. Army Corps of Engineers. "Transformation of lands at The Chesapeake & Delaware Canal, Delaware and Maryland," Philadelphia District, Philadelphia, PA. 1-1 to 1-15 and II-1 to II-20 U.S. Army Corps of Engineers (1987).

U.S. Department on Agriculture. "Soil and Water Resource Conservation Act: Appraisal 80,— Review Draft, U.S. Department of Agriculture (1980).

U.S. Department of Agriculture. "Pennsylvania Department of Agriculture, Crop Reporting Service, Pennsylvania crop and livestock annual summary 1981," U.S. Department of Agriculture, Harrisburg, PA (1981).

U.S. Department of Interior. "Surface Mining and Our Environment," U.S. Department of Interior, Washington DC (1967) p. 124.

U.S. Environmental Protection Agency. "Environmental Assessment of Reclamation of Disturbed Lands Using Wastewater Treatment Sludge," U.S. Environmental Protection Agency, Region III, Philadelphia, PA (1990).

U.S. Environmental Protection Agency. "Process Design Manual for Land Application of Municipal Sludge," U.S. EPA Report-625/1-83-016, Municipal Environmental Research Lab, Cincinnati, OH (1983).

U.S. Environmental Protection Agency. "Land Application of Municipal Sewage Sludge for the Production of Fruits and Vegetables - A Statement of Federal Policy and Guidance," U.S. Environmental Protection Agency, U.S. Food and Drug Administration, U.S. Department of Agriculture, Washington, DC (1981), p. 22.

U.S. Environmental Protection Agency. Municipal Sludge Management: Environmental Factors. Tech. Bulletin, U.S. EPA Report-430/9-76-004, MCD-28 (1977).

U.S. Environmental Protection Agency. "Environmental impact statement: Criteria for classification of solid waste disposal facilities and practices," U.S. EPA SW-821 (1979).

U.S. Environmental Protection Agency. "National Primary Drinking Water Regulations," U.S. EPA, Federal Register 40 CFR 141.11–141.16 (1979b).

Underwood, E. J. *Trace elements in human and animal nutrition,* 3rd ed. (New York: Academic Press, 1971).

University of Georgia. "Plant Analysis Handbook for Georgia, 42," Bulletin No. 735, University of Georgia, Athens (1979).

Urie, D. H., C. K. Losche, and F. D. McBride. "Leachate quality in acid mine-spoil columns and field plots treated with municipal sewage sludge," in *Land Reclamation and Biomass Production with Municipal Wastewater and Sludge,* W. E. Sopper, E. M. Seaker, and R. K. Bastian, Eds. (University Park, PA: The Pennsylvania State University Press, 1982), pp. 386–398.

Van Hook, R. I. "Cadmium, lead, and zinc distributions between earth worms and soils: potentials for biological accumulation," *Bull. Environ. Contaim. Toxicol.* 12:509–512 (1974).

Varanka, M. W., Z. M. Zablocki, and T. D. Hinesly. "The effect of digestor sludge on soil biological activity," *J. Water. Pollut. Control Fed.* 48:1728–1740 (1976).

Visser, S. "Management of microbial processes in surface mined land reclamation in western Canada," in *Soil reclamation processes*, R. L. Tate III and D. A. Klein, Eds. (New York: Marcel Dekker, 1985), pp. 203–341.

Visser, S., C. L. Griffiths, and D. Parkinson. "Effects of mining on the microbiology of a prairie site in Alberta, Canada," *Can. J. Soil Sci.* 63:177–189 (1983).

Voos, G. and B. R. Sabey. "Nitrogen mineralization in sewage sludge-amended coal mine spoil and topsoils," *J. Environ. Qual.* 16:231–237 (1987).

Water Environment Federation. "Standards for the use and disposal of sewage sludge", Final Rule. Arlington, VA. (1993). p. 204.

Water Environment Federation. 1993. "Standards for the use and disposal of sewage sludge," Final Rule. 601 Wythe Street, Alexandria, VA 22314.

Werner, W, H. W. Scherer, and F. Reinartz. "Experiences of the usage of heavy amounts of sewage sludge for reclaiming opencast mining areas and amelioration of very steep stony vineyards," in *Alternative Uses for Sewage Sludge*, J. E. Hall, Ed. (Oxford, U.K.: Pergamon Press, 1990), pp. 71–82.

Wilson, H. A. "The microbiology of strip mine spoil," West Virginia University Agriculture Exp. Stn. Bull. 506T (1965).

Wolnick, K. A., F. L. Fricke, S. G. Capar, G. L. Braude, M. W. Meyer, R. D. Satzger and E. Bonnin. "Elements in major raw agricultural crops in the United States. 1. Cadmium and lead in lettuce, peanuts, potatoes, soybeans, sweet corn, and wheat," *J. Agric. Food Chem.* 31:1240–1244 (1983a).

Wolnick, K. A., F. L. Frike, S. G. Capar, G. L. Braude, M. W. Meyer, R. D. Satzger, and R. W. Kuennin. "Elements in major raw agricultural crops in the United States. 2. Other elements in lettuce, peanuts, potatoes, soybeans, sweet corn, and wheat," *J. Agric. Food Chem.* 31:1244–1249 (1983b).

Wolnick, K. A., F. L. Frike, S. G. Capar, G. L. Braude, M. W. Meyer, R. D. Satzger and R. W. Kuennin. "Elements in major raw agricultural crops in the United States. 3. Cadmium, lead, and eleven other elements in carrots, field corn, onions, rice, spinach, and tomatoes," *J. Agric. Food Chem.* 33:807–811 (1985).

Yopp, J. H., W. E. Schmid, and R. W. Holst. "Determination of maximum permissible levels of selected chemicals that exert toxic effects on plants of economic importance in Illinois," Illinois Institute for Environmental Quality, IIEQ Doc. No. 74-33 (1974), p. 272.

Younos, T. M., Ed. *Land Application of Wastewater Sludge*, (New York: American Society of Civil Engineering, (1987).

Younos, T. M. and M. D. Smolen. "Simulation of infiltration in a sewage sludge amended mine soil," Proc. Symp. on Surface Mining Hydrol., Sedimentol., and Reclamation, University of Kentucky, Lexington (1981), pp. 319–324.

CHAPTER 5

Appendix

Table A-1. List of Common Names and Scientific Names of Vegetation

Common Name	Scientific Name
Wheat	*Triticum aestivum*
Oats	*Avena sativa*
Canada Bluegrass	*Poa compressa*
Red Clover	*Trifolium pratense*
Smooth Bromegrass	*Bromus inermis*
Alfalfa	*Medicago sativa*
Western Wheatgrass	*Agropyron smithii*
Alsike Clover	*Trifolium hybridum*
Barley	*Hordeum vulgare*
Japanese Millet	*Echinochola crusgalli var. frumentacea*
Tall Fescue	*Festuca arundinacea*
Orchardgrass	*Dactylis glomerata*
Birdsfoot Trefoil	*Lotus corniculatus*
Kleingrass	*Panicum coloratum*
Switchgrass	*Panicum virgatum*
Fescue	*Festuca megalura*
Perennial Rye	*Lolium perenne*
Bermudagrass	*Cynodon dactylon*
Reed Canarygrass	*Phalaris arundinacea*
Weeping Lovegrass	*Eragrostis curvula*
Redtop	*Agrostis gigantea*
Deertongue	*Panicum clandestinum*
Blue Gamma	*Boutelova gracilis*
Sweetgum	*Liquidamber styraciflua*
Sitka Spruce	*Picea sitchensis*
Galleta	*Hilaria Jamesii*
Bottlebrush Squirreltail	*Sitanion hystrix*
Fourwing Saltbush	*Atriplex canescens*
Mountain Big Sagebrush	*Artemisia tridentata*
Ladino Clover	*Trifolium repens*
Panic Grass	*Panicum dichotomiflorum*
Sudan Grass	*Sorghum vulgare sudanenese*
Serecia Lespedeza	*Lespedeza cuneata*
Crownvetch	*Coronilla varia*

Table A-1.   List of Common Names and Scientific Names of Vegetation (Continued)

Common Name	Scientific Name
Annual Ryegrass	*Lolium multiflorum*
Sideoats Gramagrass	*Bouteloua curtipendula*
Canada Wild Rye	*Elymus canadensis*
Foxtail Grass	*Setaria* spp.
Korean Lespedeza	*Lespedeza stipulacea*
Sweetclover	*Melilotus* spp.
Kobe Lespedeza	*Lespedeza striata*
Slender Wheatgrass	*Agropyron trachycaulum*
Intermediate Wheatgrass	*Agropyron intermedium*
Pubescent Wheatgrass	*Agropyron trichophorum*
Crested Wheatgrass	*Agropyron desertorum*
Meadow Brome	*Bromus erectus*
Timothy	*Phleum pratense*
Corn	*Zea mays*
Soybean	*Glycine max*
Highbush Blueberry	*Vaccinium corymbosum*
Bush Bean	*Phaseolus vulgaris*
Virginia Pine	*Pinus virginiana*
Hybrid Poplar	*Populus* spp.
Black Locust	*Robinia pseudoacacia*
European Alder	*Alnus rugosa*
Red Oak	*Quercus rubra*
Cottonwood	*Populus deltoides*
Eastern White Pine	*Pinus strobus*
Silver Maple	*Acer saccharinum*
Green Ash	*Fraxinus pennsylvanica*
Loblolly Pine	*Pinus taeda*
Red Pine	*Pinus resinosa*
White Pine	*Pinus strobus*
Austrian Pine	*Pinus nigra*
White Spruce	*Picea glauca*
Japanese Larch	*Larix leptolepis*
Black Walnut	*Juglans nigra*
Bristly Locust	*Ribinia fertilis "Arnot"*
Galleta	*Hilaria jamesii*
Bottlebrush Squirreltail	*Sitanion hystrix*
Fourwing Salt Bush	*Atriplex canescens*
Mountain Big Sagebrush	*Artemisia tridentata*
Kentucky Bluegrass	*Poa Pratensis*
Red Fescue	*Festuca rubra*

Index

Acid spoils, pH of, 93–94
Actinomycetes, in soil, 108, 121
Aerobic heterotrophic bacteria, in soil, 106, 121
Agronomic crops, *see* Field crops
Anaerobically digested sludge, grass and legume growth and, 29–30
Animal health, 132–139
  nitrates and, 52
Anthracite, *see* Coal mining; Coal refuse material

Bacteria
  heterotrophic, aerobic, 106, 121
  nitrifying, 108–109, 121–122
Beans, growth of, 37, 39
Biological properties of soil, 103–126
  Actinomycetes and, 108, 121
  bacteria and
    aerobic heterotrophic, 106, 121
    nitrifying, 108–109, 121–122
  community respiration, 109–110
    organic matter and, 122
  earthworms and, 125
  fungi and, 106, 108, 121
  microbial decomposition, 110–111, 123
  nitrogen mineralization, 124
  organic matter and, 114–124
    grass and legume growth and, 28
  pathogenic organisms and, animal health and, 137–139
  trace metals and, 112–114
Biomass, root, soil organic matter and, 117–118
Birds, trace metals in, 133

Birdsfoot trefoil, *see also* Grass species
  macronutrients in, 49, 69, 71
  trace metals in, 66–80
Blackbirds, trace metals in, 133
Blueberries, growth of, 37, 40
Blue Mountain, 6
Boron, plant concentrations of
  grass species, 64, 66, 69
  tree species, 84, 86
Bray phosphorus, soil concentrations of, 101

Cadmium, *see also* Trace metals
  plant concentrations of
    grass species, 56, 58, 64, 66, 67, 74, 76, 81
    tree species, 84–86
Calcium levels
  grass species and, 63
  soil, 101
Carbon, organic, soil content of, 119
Cation exchange capacity (CEC), 94
Chemical properties of soil, 92–103
Chromium, *see also* Trace metals
  plant concentrations of, grass species, 64, 66, 68, 79
Coal mining, 2, *see also* Mining
  land disturbed by, status of, 1
  refuse banks and, 5–6, *see also* Coal refuse material
  trace metals and, 66–80, 96
Coal refuse material
  soil pH and, 100–101
  trace metal uptake by grass and, 59
Cobalt, *see also* Trace metals
  plant concentrations of, grass species, 64, 66, 67, 77

Colorado, land reclamation projects in, 16
Community respiration, soil, 109–110
  organic matter and, 122
Composite root mass, soil organic matter
    and, 117–118
Composted sludge, grass and legume growth
    and, 25–26
Conductivity, electrical, soluble salt content
    of soil and, 92–93
Conifer species, growth of, 41–47
Copper, see also Trace metals
  plant concentrations of
    grass species, 56–57, 64–66, 74, 77, 81, 86
    tree species, 84–86
Corn
  nutrients in, macronutrients, 40, 48
  yields of, 37, 39, 40
Cottontail rabbits, trace metals in, 133–136
Crops, see Field crops
Crownvetch, see also Grass species
  macronutrients and, 69, 71
  trace metals in, 56–57, 61–66
    deep coal mine refuse bank and, 66–80

Decomposition
  microbial, soil and, 110–111, 123
  organic matter, soil and, 116–117
Delaware, land reclamation projects in, 19
Dewatered heat-dried sludge, trace metal
    concentrations and, 96, 97
Disturbed land, status of, 1–6
Drinking water, see Water quality
Dry matter production, 32–33, 35–37, 116
  trace metals and, 62, 74

Earthworms, 125
Electrical conductivity (EC), soluble salt
    content of soil and, 92–93
Emissions, land disturbed by, 6
Environmental Protection Agency (EPA)
  40 CFR Final Rules for Use and Disposal
    of Sewage Sludge, 139–140
  guidelines and regulations of, 8–9
  trace metals in plants and, 78–79
EPA, see Environmental Protection Agency

Federal regulations, sludge on mine land and,
    8–11
Fertilizers
  inorganic, sludge versus, 29
  root biomass and, 117–118
Fescue, see also Grass species
  growth of, 23, 29
  macronutrients in, 49

Field crops, see also Vegetation; specific
    crops
  growth responses of, 37, 39–40
  quality of, trace metals and, 88–90
Final Rules for Use and Disposal of Sewage
    Sludge, 139–140
Fly ash, sludge mixed with
  grass species growth and, 33–37
  grass trace metal concentrations and,
    59–61
  soil chemical properties and, 96–97
  tree growth and, 45–47
  tree trace metal concentrations and, 84–87
Forage grasses, see also Grass species
  growth of, 29
  trace metal uptake by, coal refuse material
    and, 59
Fourwing saltbush
  growth responses of, 43–44
  trace metals in, 83
Fungi, in soil, 106, 108, 121

Georgia, land reclamation projects in, 16
Germany, land reclamation project in, 21
Grass species
  growth responses of, 13, 22–39
  quality of
    macronutrients and, 48–51, 69, 71
    trace metals and, 52–81, 86–87
Grazing animals, 132–133
Groundwater
  nitrate–nitrogen in, 126–128
  trace metals in, 76, 83, 128–131
Growth, of vegetation
  field crops, 37, 39–40
  grass and legume species, 13, 22–39
  trees, 41–47
Gypsum, grass growth and, 30

Hardwood species, growth of, 41–47
Herbaceous vegetation, see also Vegetation
  growth of, 33, 35–37
  macronutrients in, 51
  trace metals in, 60, 61
Heterotrophic bacteria, aerobic, in soil, 106,
    121
Highbush blueberries, growth of, 40

Illinois, land reclamation projects in, 16–18
Inorganic fertilizers, sludge versus, 29
Iron ore tailings revegetation, 29

Kentucky, land reclamation projects in, 17
Kjedahl nitrogen, soil concentrations of, 101

Lagoon dried sludge, grass and legume growth and, 28–29
Land reclamation projects, data on, 14–21
Lead, *see also* Trace metals
  plant concentrations of
    grass species, 64–66, 74, 75, 81
    tree species, 84–86
Lead smelters, emissions from, 6
Legume species
  growth responses of, 13, 22–39
  quality of, trace metals and, 52–81, 86–87
Lime treatment
  grass growth and, 30–32
  soil metal concentrations and, 102–103
  soil pH and, 93–94
Liquid sludge
  digested, tree species and, 87–88
  undigested, tree species and, 87
Livestock, grazing of, 132–133

Macronutrients, *see also* Nutrient concentrations
  vegetation quality and, 48–52, 63–64, 69, 71
Magnesium levels
  grass species and, 63
  soil, 101
Manganese
  plant accumulation of, 56, 78
  plant concentrations of, tree species, 84–86
Maryland, land reclamation projects in, 19
Meadow voles, trace metals in, 133, 137, 138
Metal concentrations, *see* Trace metals
Microbial decomposition, 110–111, 123, *see also* Biological properties of soil
Mineralization, nitrogen, 124
Mining, *see also* Coal mining
  land disturbed by
    status of, 1–6
    vegetation growth on, 13, 22–47
  sludge on land used for, regulations governing use of, 8–11
Mountain big sagebrush
  growth responses of, 43–44
  trace metals in, 83

New Mexico, land reclamation project in, 19
Nickel, *see also* Trace metals
  plant concentrations of
    grass species, 64, 66, 74
    tree species, 84–86
Nitrate levels
  groundwater, 126–128
  soil water, 126–128
  toxicity to animals and, 52

Nitrifying bacteria, in soil, 108–109, 121–122
Nitrogen levels, 48–49, *see also* Nutrient concentrations
  grass species and, 63
  groundwater, 126–128
  Kjedahl nitrogen in soil, 101
  soil, 118–119
    carbon ratio to, 119
  soil water, 126–128
  toxicity to animals and, 52
Nitrogen mineralization, 124
Nutrient concentrations
  field crop, 40
  grass species, 48–52, 63–64
  tree, 45

Office of Surface Mining Reclamation and Enforcement, 1
Ohio, land reclamation projects in, 19
Oklahoma, land reclamation project in, 19
Orchardgrass, *see also* Grass species
  trace metals in, 61–66
Organic matter
  dead
    accumulation of, 116
    decomposition of, 116–117
  soil, 114–124
    grass and legume growth and, 28

Palmerton project, *see* Fly ash, sludge mixed with
Panic grass, *see also* Grass species
  trace metals in, 72, 80
Pathogens, animal health and, 137–139
Pennsylvania, land reclamation projects in, 14–16, *see also* Fly ash, sludge mixed with
Percolate water, trace metal concentrations in, 75, 82
pH, soil, 93–94, 100–101
Pheasants, trace metals in, 133
Philadelphia Water Department program, trace metals and, 73–76, 81–83
Phosphorus levels, 48–49, *see also* Nutrient concentrations
  Bray phosphorus in soil, 101
  grass species and, 63
Physical properties of soil, 91–92
Phytotoxicity, manganese and, 56
Pine, growth of, 41–47
Plants, *see* Vegetation
Potassium levels
  grass species and, 63
  soil, 101

Potassium treatment, grass growth and, 31–32

Rabbits, trace metals in, 133–136
Redtop, *see also* Grass species
 growth of, 29
Red-winged blackbirds, trace metals in, 133
Reed canarygrass, *see also* Grass species
 growth of, 29
 trace metals in, 56–57
Refuse banks, coal mining, 5–6, *see also*
  Coal refuse material
 trace metals and, 66–80, 96
Regulations, sludge on mine land and, 8–11
Respiration, community, soil, 109–110, 122
Revegetation, *see also* Vegetation
 iron ore tailings, 29
Rock phosphate additions, vegetation growth
  and, 49
Root mass, soil organic matter and, 117–118

Sagebrush
 growth responses of, 43–44
 trace metals in, 83
Salt, soluble, soil content of, 92
Saltbush
 growth responses of, 43–44
 trace metals in, 83
Sludge use on mine land, *see also* Mining
 regulations governing use of, 8–11
Smelters, emissions from, 6
Soil, 91–126
 Actinomycetes in, 108
 bacteria in
  aerobic heterotrophic, 106, 121
  nitrifying, 108–109, 121–122
 biological properties of, 103–126
 chemical properties of, 92–103
 community respiration in, 109–110
  organic matter and, 122
 fungi in, 106, 108
 horizon development, organic matter and,
   118
 microbial decomposition in, 110–111, 123
 organic matter in, 114–124
  grass and legume growth and, 28
 pH of, 93–94, 100–101
 physical properties of, 91–92
 trace metals in, 96–100, 112–114
Soil water
 nitrate–nitrogen in, 126–128
 trace metals in, 128–131
Soluble salt, soil content of, 92
South Carolina, land reclamation project in,
  20

Soybean yields, 37, 39
State regulations, sludge on mine land and,
  8–11
Surface mining, *see also* Mining
 land disturbed by, status of, 1
Surface Mining Control and Reclamation Act
  of 1977, 1, 6–8
Surface water, quality of, 131–132
Sweetcorn grain, macronutrients in, 48
Swine, trace metals in, 133

Tall fescue, *see* Fescue
Tennessee, land reclamation project in, 16
Texas, land reclamation projects in, 19
Toxicity to animals, *see also* Animal health
 nitrogen and, 52
Trace metals
 animal health and, 132–139
 decreasing concentrations of, over time, 55
 soil content of, 96–100, 112–114
 vegetation quality and, 52–90
  field crops, 88–90
  grass and legume species, 52–80, 81
  grazing animals and, 132–133
  trees, 80, 82–88
  water quality and, 128–131
Tree species
 growth responses of, 41–47
  amount of sludge and, 42
 quality of
  macronutrients and, 52
  trace metals and, 80, 82–88

Underground mining, *see also* Mining
 land disturbed by, status of, 1
United Kingdom, land reclamation projects
  in, 20–21
U.S. EPA, *see* Environmental Protection
  Agency (EPA)

Vegetation
 common and scientific names of, 157–158
 growth responses of, 13, 22–47
  field crops, 37, 39–40
  grass and legume species, 13, 22–39
  trees, 41–47
 quality of, 48–91
  field crops, 88–90
  grass and legume species, 52–81
  macronutrients and, 48–52, 69, 71
  trace metals and, 52–90
  trees, 80, 82–88
Virginia, land reclamation projects in, 14,
  19–20

Water
  groundwater
    nitrate–nitrogen in, 126–128
    trace metals in, 76, 83, 128–131
  percolate, trace metals in, 75, 82
  soil
    nitrate–nitrogen in, 126–128
    trace metals in, 128–131
Water movement, vegetative cover and, soil
        organic matter and, 116
Water quality, 126–132
  soil and groundwater, 126–131
  surface water, 131–132

West Virginia, land reclamation projects in, 20
Wildlife habitat, 23
Wisconsin, land reclamation projects in, 20
Worms, 125

Zinc, *see also* Trace metals
  plant concentrations of
    grass species, 56, 58, 64, 66, 73, 81
    tree species, 84–86
Zinc smelters
  emissions from, 6
  soil pH and, 94